Nikola Tesla's
Wireless Power Secrets Revealed

The Next Generation of Tesla's Wireless Power Systems

Reactive Power Transport and Broadband Data

By G. Martin Poole

Nikola Tesla The Next Generation Secrets Revealed

Copyright © 2019 All Rights Reserved

Published by Amazon Inc.

United States of America

Important insights from the Author
- Digging deeper into Tesla's wireless power system

Tesla is an interesting person to study and though he died on January 7, 1943 his life's work lives on. He was complicated, sensitive, an amazing engineer, and a scientist extraordinaire. After reading his autobiography, patents, and the many stories about him, it seems sometimes that I knew the man personally. At times, while researching his patents and writings, I would find myself asking him what the heck did he mean by that math calculation. Of course, I expected no answer, but I surely asked many questions. At times, I would find answers to my questions in unusual places such as in interviews he did with newspaper or magazine publishers, or articles that he wrote for public consumption. In the trove of records available on Tesla, sometimes only a brief hint or the mention of a subject is all that exists.

Tesla made it a habit to reveal an idea but was purposefully vague in the patent design, giving only enough information to make a rough working model. This was especially common in his wireless work. One example is the capacitor symbol across a primary coil symbolizing an undamped, spark-gap system. [1]A letter 'G' symbolized Generator, and 'S' referred to an electrical power-out or supply source. In all his patents he was careful to provide a working design, but it was a very inefficient one if taken in exact context.

I have written this book to tell of the rediscoveries our company (Wireless Power Technologies, USA) has made in our research of the Tesla resonant, Reactive Power Transport (*RPT*) system. Nothing is withheld from the reader about how the wireless power transfer and communication operate through the earth. There is no other publication where this information is available. It stems from documented discoveries made during our research. Included here are illustrations and mathematical proofs which apply standard electrical engineering terms, formulae, and Maxwell's equation sets.

---

[1] US patent 1119732 Nikola Tesla 1907

Rediscoveries:

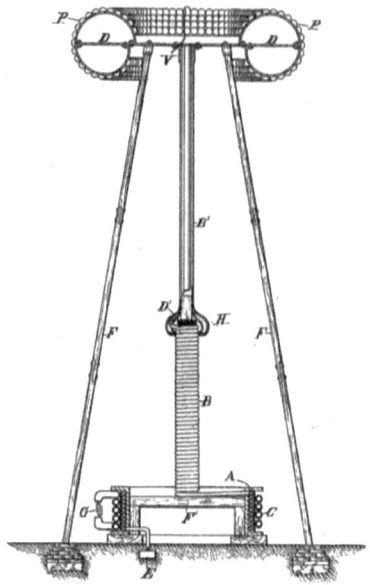

A Tesla drawing depicting a wireless transmitter

At the point of writing this book I have devoted six years of study and research in Tesla's wireless technology. Various working models and functioning prototypes of wireless power arrangements were built and tested. Initiatives stemming from our research are:

- light speed underwater Internet/broadband with power transfer

- aerial power transfer with communication for drones or cars

- resonant wireless power transfer through the earth

- broadcast only systems (like radio)

- Wireless Power Broadcast

Nothing is held back between these covers about the workings of Tesla's wireless system. This book is specifically written to reveal the technological details of wireless power transport. The book will detail the longitudinal electron wave (*LEW*), the electron plasma field (*EP*), and the magnified electrical power in standing, resonant compression waves. These topics and more will be covered in detail with mathematical proofs, foot note references, glossary, Footnote links to important WEB information, and a bibliography at the end of the book. Experiments and conclusions in this book are accompanied with scientific proofs that meld existing science and connect current theory with the art of power transfer à la Tesla. New terminology is introduced here that can be used to construct discourse and a vocabulary for the development of this discipline.

The simplicity of the wireless-power-transfer art will leave you wondering how we missed it in our studies of Electrical Engineering.

Table of Contents

Chapter 1: The first Wireless Power

Tesla had completed his work with rotating magnetic field motors and the Niagara power plant in the early eighteen-nineties. He then began in earnest to work with high frequency currents for wireless power transmission. Tesla learned by experience that physical plant construction for power transport was expensive. The poles and power lines from Niagara to Buffalo, NY was an expensive venture needing full-time employees for maintenance. The high cost threatened deployment of AC power to the rest of the country. Tesla's intent was to develop a long-haul power transport method to reduce as many of the wires, transformers, and poles as possible.

Tesla's first tries at wireless power transmission were in electrostatics. His simple scheme was specifically for indoor lighting of houses and businesses. To develop the idea, Tesla made a high frequency multi-pole motor generator, see figure 1.4. This machine created sine waves up to 10 kHz (*later a 20 kHz one*) at 10 Kw electricity powering his resonant, electrostatic transmitter plates in a damped or undamped manner.

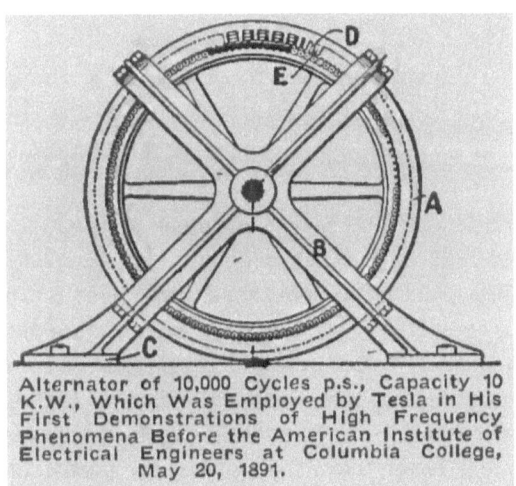

Alternator of 10,000 Cycles p.s., Capacity 10 K.W., Which Was Employed by Tesla in His First Demonstrations of High Frequency Phenomena Before the American Institute of Electrical Engineers at Columbia College, May 20, 1891.

Figure 1.4 The Multi-pole Generator

The generator connected directly to the primary coil of a transformer with a variable capacitor placed in parallel. Efficiency increased through tuning with a variable capacitor. The generator could work from 10 Hz to 10,000 Hz. Tesla designed a special high frequency air core transformer immersed in oil (*to prevent damage caused by high voltage arcing*) and connected its secondary to one or two large metal plates.

The one plate, one wire method was unusual because, unlike normal current flow through a loop, this apparatus did not need a closed circuit or two capacitor plates. Figure 1.5 represents a reactive arrangement and consumes little or no power from the source to operate. The power induced into the wire by the secondary fully charges the terminal plate and then discharges it in an oscillating fashion. Simply, Tesla was charging and discharging a one plate capacitor and using the electrostatic field to light [2]vacuum bulbs. One plate power transfer proved that electricity is transportable over one wire and is consumed from the electric 'E' field emanating from the plate. It was this principal motif that Tesla applied to many of his wireless power transfer schemes.

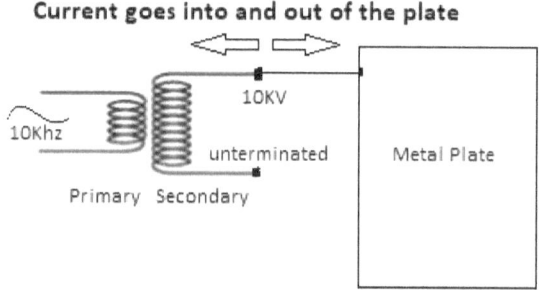

Figure 1.5 A Reactive, electrostatic wireless power system

Other types of Tesla's experiments with wireless power transfer performed in the late 1880s', used high frequency electricity, with two electrostatic plates. In effect, Tesla placed two capacitor plates about ten feet apart, and set the high frequency generator to work. Figure 1.6 is a picture of the experiment shown at the Electrical Engineer's Conference in 1891. In Tesla's setup, metal plates were hung with non-conductive cords, and connected to the secondaries of a high-voltage transformer. In his demonstration, Tesla would hold vacuum tube lamps in his hand as he stood between the charged plates, and they would light brightly. Tesla himself was a third plate[3], in this unusual arrangement.

---

[2] https://teslaresearch.jimdo.com/articles-interviews/tesla-bulbs-the-electrical-experimenter-june-1919-volume-vii-no-74/

[3] C3 (3 plate) capacitors are used by Wireless Power Technologies to reactively receive signals through the NM

[4]Figure 1.6 Wireless Power Display at IEEE Columbia College in May 1891

He continued experimentation with this electrostatic form of wireless power and developed various lamps which created bright, natural light. The inventions flowing from this research were fluorescent, neon, and vacuum tube lightbulbs, all with no filaments. Tesla installed this style lighting in his laboratory; he could pick up a bulb and move it anywhere in the room for light. The vacuum tube lighting system never publicly caught on at the time because Edison's DC powered lamps had a lock on the consumer market. Neither could Tesla use a standard, patented screw-in base on his AC lamps without paying a steep royalty fee to Edison. Tesla used his vacuum-tube lamps and Niagara Falls power at the [5]1893 Columbia Exposition to light the buildings and walkways displaying his superior high-lumen, low-power lighting.

During his light bulb making experiments, Tesla discovered the [6]vacuum tube capacitor, see figure 1.7. The capacitance was much greater than a plain metal sphere many times the size.

---

[4] Tesla memorial society of New York
[5] https://teslasciencecenter.org/pivotalmoments/columbian-expositions/
[6] These are not found anywhere in the market and must be custom made. They died with Tesla, but may have been used in the Hotel New Yorker on the roof.
https://www.teslasociety.com/hotelnewyorkerroom.htm

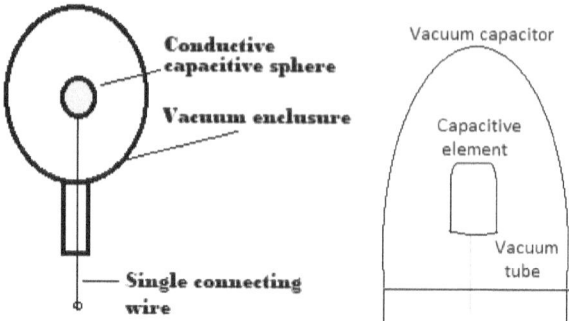

Figure 1.7 The Single Plate Vacuum Tube Capacitor

The principle of Tesla's vacuum tube capacitors is still cutting-edge technology in high voltage, high frequency radio. Tesla hired a German glass blower to make these capacitors with plug-in bases to replace the large metal copula of his Wardenclyff, wireless power transmitter to lessen the cost, weight, and wind load. The glassblower is the same person who made his vacuum tube lamps used in the [7]1 893 Columbia Exposition. This explains the various designs of the tower's cupola, some show a solid metal top (*concept design*), and others have the skeletal look of vacuum tubes. The vacuum tubes cut the cost of the tower and replaced the expensive, high maintenance, curved metal cupola.

The vacuum tube terminal capacitor design offered more electrical capacitance and replaced the solid metal cupola in the original design. Added benefits of the vacuum tube were: it was weatherproofed, supported higher voltages, lower wind loading, and prevented arcing in rain storms. Vacuum capacitors greatly reduced the height requirement and lessened atmospheric power losses. The vacuum tubes also prevented upper atmospheric power conduction. Tesla's vacuum tube capacitors placed on tall buildings could easily convert them into a resonant power transmitter or receiver for metropolitan power transport.

The knowledge Tesla gained from experiments with electrostatics defined his world-wide power transmission development. Experiments with the one wire high voltage transformer arrangements proved how one conductor could carry power. Later after several patents on single wire schemes, Tesla explains that he had the 'happy' idea of using the earth as the 'one' conductor instead of a wire or the atmosphere.

Tesla earnestly began his effort at wireless power transmission in New York City where he had a laboratory on Fifth Avenue. It was there that Tesla did most of his research on wireless power transmission until fire destroyed the lab under suspicious circumstances in March, 1895. Tesla had powerful enemies, and he always thought the fire was intentional. The lab destruction set him back at least

---

[7] https://www.britannica.com/event/Worlds-Columbian-Exposition

two years, but Tesla stoically resumed his work as soon as possible at a new lab on Houston Street in New York. There was no insurance to cover the loss at his Fifth avenue lab. He lost a fortune in one-of-a-kind equipment developed for his wireless work.

Figure 1.7 The site of the Houston Street Lab from *NY city subway archive*

By 1893, Tesla had theorized how power transfer through the earth could be possible. By the mid eighteen-nineties he had already performed many experiments in New York using earth as the transmission medium. Details of these experiments are in Anderson's book_On His Work with Alternating Currents_. A loyal friend who owned the Colorado Springs power plant invited him there, and his plan to scale up his transmitter began. The rural setting provided a location where there was negligible grounding noise to interfere in his tests. In the city, there were motors, streetcars, and various power returns which made it difficult to ferret out his own earth signals. The Colorado Springs power station owner had promised Tesla free power for his wireless power experiments. He later said that he felt a warm welcome when he arrived there.

[8]The mines in Colorado Springs used Tesla's AC (*Westinghouse*) motors to drive their equipment, and the power company used his patented generators to provide commercial power to the city. Tesla had already made construction plans for his laboratory near Pike's Peak, and bought the land to build it. The book, Colorado Springs Notes by Tesla, describes the experimentation and the challenges he faced there. Tesla was indeed a man on a mission: his personal spending for shipping, lumber, equipment, and manpower proved that.

Tesla wrote the paper, "The Problem of Increasing Human Energy" for Century Magazine in 1900 after returning from Colorado Springs. J.P. Morgan read the paper and became interested in investing in Tesla's enterprise. After meeting with him a few times, Morgan decided to finance Tesla's tower at Wardenclyff on Long Island, NY for an original grant of $150,000. Tesla got underway building it, but he badly underestimated the cost. At the time of the construction, a financial recession triggered drastic price increases in the materials he needed. When Tesla went back to Morgan for more funds to complete the project, Morgan outright refused him.

This is all part of history, but no one knows the exact reasons for Morgan's attitude change toward Tesla. An interesting point is Thomas Edison, who loathed Tesla, was a major business partner of Morgan. The gossip of the time was when Marconi beat Tesla to the punch with radio transmission across the Atlantic, J. P. Morgan lost interest in Tesla's project, and then abandoned him. Another rumor said Tesla did not plan to charge users for the power from the Wardenclyff tower, and there was no metering plan in place to do so. This stirred Morgan to proclaim that he wanted no part of giving away power. Strangely, of the three planned Wardenclyff oscillators the communication tower is the only design carried out. It could deliver power, but on a small-scale compared to a large dedicated power tower.

The ironic rub in the Tesla and Marconi 'WHO WAS FIRST' radio transmission fiasco was the fact that Tesla had already built the device able to [9]broadcasts across the Atlantic and tested it, but he did not document or divulge it to the public. Tesla lost out to Marconi, and Marconi stole his ideas to do it; a patent infringement lawsuit filed against Marconi by Tesla followed; Tesla posthumously won the law suit, but Marconi got the glory. There were no textbook corrections made to resolve the injustice toward Tesla and his radio patents. Look up Marconi in a science textbook or online and you can see for yourself.

Tesla looked for other sponsors to complete Wardenclyff, but he was a 'hot potato' because of Morgan's influence in corporate America. Tesla's reclusive

---

[8] Tesla's Colorado Spring Notes 1899
[9] On His Work with Alternating Current pages 83, 84, 85, 106

lifestyle and sarcasm did not help matters either. The bankruptcy of Tesla's company, and the loss of the Wardenclyff power station on Long Island later followed. Finally, I should mention Tesla's attitude about billing for electric power usage.

The power in the transmission wave was needed for the communication link. Tesla reasoned, "How could he charge others for it?". The longitudinal standing wave resonance power (*more on this in chapter four*) magnifies with each wave and will go out control if not used. The excess power would damage his transmitter and receiver equipment. He was content to just sell the communication services, and give away the power. This is unlike the oil and mining companies. They surely did not hesitate invoke a big-money business model on natural resources. Morgan was right in that regard. Tesla was as poor a business man as he was a brilliant engineer.

Chapter 2: Tesla's methods for Wireless Power Transmission

It has been over one hundred and thirty years at this writing since Nikola Tesla invented and patented the machines and the modern power grid. The grid

Figure 2.0 A look inside the Niagara power plant designed by Tesla

encompassed more than twenty miles of wires and power poles stretching from Niagara Falls to Buffalo, NY. This was the first commercial AC power grid in the U.S. Construction of the Hydroelectric plant began in 1888 and completed in 1895. Buffalo was electrified and the grid expanded, eventually to the whole northeastern U.S.A.

This was the first step toward providing AC electricity to the world. Tesla realized early on that transporting electricity over wires was expensive. The construction costs, manpower expenses, maintenance, and various transport losses added up. Capital investment required profits and the margin was slim. In Tesla's reasoning there would be no electric power for the world's people until a better way was found for transport. He began his research into ways to do it wirelessly. And, toward that effort, Tesla set to work on the challenge, began to test various solutions, and to apply for patents on the workable ones.

The original patented plan for Tesla's Wireless Power Transfer was not practical but it would work. The plan was to use large Hydrogen balloons at thirty thousand feet with conductive wires attached to power transformers on the ground. Figure 2.1 shows the patent application. It matched the actual plan shown in figure 2.2. Evacuated, conductive tubes mimicked the rarified Ionospheric conductive properties. The final working model used balloons with wires to connect the transmitter and receiver through the conductive Ionosphere and earth ground plates. The power supply was a high-frequency generator connected directly to the Ionosphere and earth ground.

A high frequency, resonant-signal lock at ten thousand Hz. with a high voltage (*3 to 5 million volts*) directed to the Ionosphere via the transmitter's balloon connection obtained resonance with the receiving station. The receiver's conductive balloon and wire provided the primary coil's connection to the power. The other terminal of the primary was connected to earth ground. The receiver's secondary winding produced the consumable Electrical power. [10]The transmitter and receiver formed a concatenated, tuned, resonant scheme with low frictional power losses.

This arrangement of components with transmitter and receiver, was a standard closed-circuit arrangement. The upper atmosphere served as the 'hot' wire and the earth as a return wire. The scheme was unlike his [11]RSWT in patent in figure 2.3. The difference being the Tesla balloon system was a closed-circuit, resonant Kirchhoff arrangement with two transport conductors, the atmosphere and the earth.

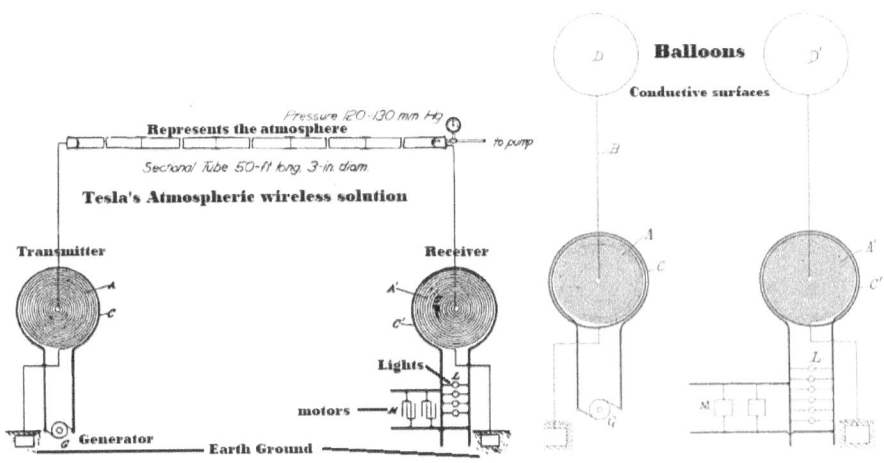

Figure 2.1 Atmospheric simulation for a patent   **Figure 2.2 Actual wireless system**

This Tesla Wireless, balloon design was to provide electricity for commercial, domestic, and rural use. The power source served customers hundreds or even thousands of miles away. The plan called for a power transmitter driven by

---

[10] ten to twenty million volts at twenty thousand Hz created by a pancake coil system
[11] https://www.youtube.com/watch?v=Qz1-RIcj1HY and
https://file.scirp.org/pdf/ENG20121100002_50196856.pdf

renewable hydroelectric energy. The natural land elevation at the dam location provided nearness to the Ionosphere and used less connecting wire to the balloons.

In Tesla's patent drawing shown in figure 2.1, the generator is labeled by two circles with a 'G' underneath. The two lines touching the generator connecting it to the primary symbolize generator output connections (*brushes*). Specific symbols at the receiving end are electric motors and lamps. Old style capacitor icons identified [12]electrostatic (*nonmagnetic*) motors, and circles labeled Tesla's own, efficient broad-spectrum vacuum tube lights. These symbols were common in his wireless patents.

Tesla's plan was to replace the balloons with a conductive, [13]particle beam to the clouds. The patent was for large plants and seagoing vessels. The first wireless power transfer patent (*figure 2.1*) hangs on the premise the atmosphere is conductive at thirty miles up. Tesla verified by experiment that the atmosphere was indeed a conductor. He tested the patent with Hydrogen balloons connected with wires to the transmitter and receiver. He proved the existence of the Ionosphere predicted by Gauss in 1839.

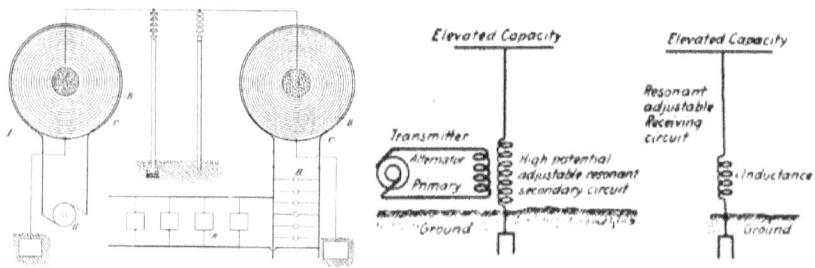

Frame A-_Wired_   **Frame B-**_Wireless-_ used at Houston St.

Figure 2.3 Basic single wire transmitter – and a wireless equivalent using earth ground

Tesla's next idea for long-range wireless power made use of the design shown in figure 2.3, frame 'A'. Tesla exchanged the overhead connecting wires with the earth ground design shown in Frame 'B'. If you recall, this motif was part of his single-plate electrostatic arrangement shown in chapter one. The single wire for connecting the two oscillators was the earth and not an aerial wire. At first, Tesla tried to use voltages from three million volts or higher to etch an electron path to the Ionosphere from the cupola. He reasoned the trickle of electrons upward would become a conductive path. The idea worked, but it lit the sky like the Aurora

---

[12] https://teslauniverse.com/nikola-tesla/articles/alternate-current-electro-static-induction-apparatus
[13] A much later version of the particle beam technology was named the 'Death Ray': it has been proposed that this was the basis of the Star Wars technology in the 1980s. It was an 'open ended' vacuum tube.

Borealis. The arrangement was inferior to the resonant reactive one because the transport medium consumed power to create light.

The copula capacitor in the later design of the wireless power transfer wasn't intended to etch an atmospheric connection upward but as electrical storage in resonant *reactive* power transfer circuit. Hertz wave radiation was prevented by the signal absorbing cupola. The higher terminal increased the capacitance by 2.7 percent each foot above earth and prevented electric arcs and shocks for anyone nearby.

The wireless design shown in 2.3- frame 'B' was used in New York at Tesla's Houston Street lab. In Colorado Springs he replaced the damped generator design with the undamped, spark gap, high frequency arrangement of figure 2.5 also developed in New York. The spark gap was the innovation Tesla used to do away with mechanical generators. The oscillator produced [14]undamped waves of any frequency. Tuning the primary section ([15]*tank circuit in the resonant section 1 of fig 2.5*) to match the wavelength of the [16]quarter wave secondary increased efficiency of the machine through multi-stage resonance. This was part of his tuned, concatenated circuits scheme. It [17]lowered electrical friction through the use of parallel and series resonant combinations in the transceivers. This was a clever solution for transporting power [18]reactively, through resonance and nearly free of cost.

The undamped spark gap method allowed Tesla to fashion the first continuous-wave high frequency electrical oscillator. Previously, multi-pole devices generated the high frequency currents and upper limits were defined by mechanics. The spark gap oscillator made generation of any frequency possible. Using simple algebra and the resonant frequency formula: $F_r = 1/2\pi\sqrt{LC}$ frequency adjustments were simple.

---

[14] Undamped: little or no attenuation. The LC resonant section 1 in fig. 2.5 would oscillate for a longer period of time after each power stroke making the transmitter more efficient and maintaining resonance
[15] A tank is an arrangement of a capacitor and an inductor in parallel or series that resonates at a planned frequency.
[16] quarter wave length – the time of travel for power to go up the antenna to the terminal establishes the quarter wavelength of the frequency. The actual frequency is found by multiplying the quarter wave by 4.
[17] Resonant circuits http://hyperphysics.phy-astr.gsu.edu/hbase/electric/serres.html
[18] https://www.techopedia.com/definition/15008/reactive-power

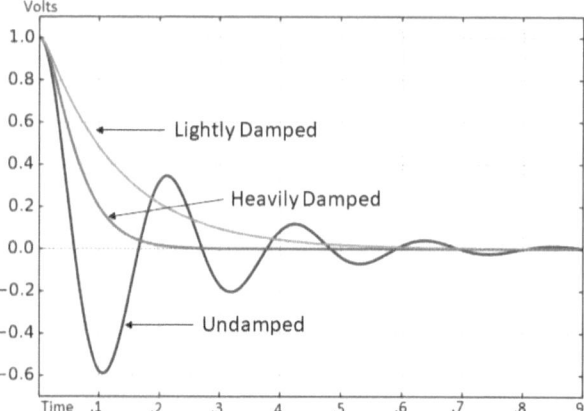

Figure 2.4 The difference in undamped and damped oscillations and the relationship of

In undamped circuits, oscillations continue for a few cycles after they are created. The spark-gap supports two functions in the Tesla oscillator. It switches off for the capacitive charging cycle then when the charge is at maximum it arcs and provides a conductive path for the [10]resonant tank circuit to oscillate. It is powered by the capacitive discharge.

Spark gaps are normally part of Tesla undamped oscillators. Oscillations are damped (*suppressed*) when friction (*resistance*) immediately consumes their power. Figure 2.4 shows the effect of dampening resistance on free oscillations. The spark gap is a temporary short circuit (*fig. 2.5*) allowing the LC resonant primary to create oscillations when it fires. The spark provides continuity, but it also short-circuits the power transformer secondary consuming power. A redesign corrected the problem. Modern technology has replaced the noisy (*audible and radio noise*) spark gap with solid-state devices or power vacuum tubes.

In figure 2.5, the earth, with a near zero resistance, connects these two halves of a four-part resonant network. Tesla referred to them as 'four tuned concatenated circuits', two tuned circuits on the transmitter and two on the receiver. Both systems resonate together through the earth between the transmitter sphere to the receiver sphere every full cycle. The series resonance of the terminals and ¼ wave coils provides frequency selectivity and power efficiency in transport. The parallel resonance in the primary offers a high resistance at resonance rendering undamped waves. Power is picked off for the load at the receiver's quarter-wave, resonant secondary. The same arrangement of resonant circuits was also used in Tesla's communication network for *individualization* (*encryption*) in modern jargon.

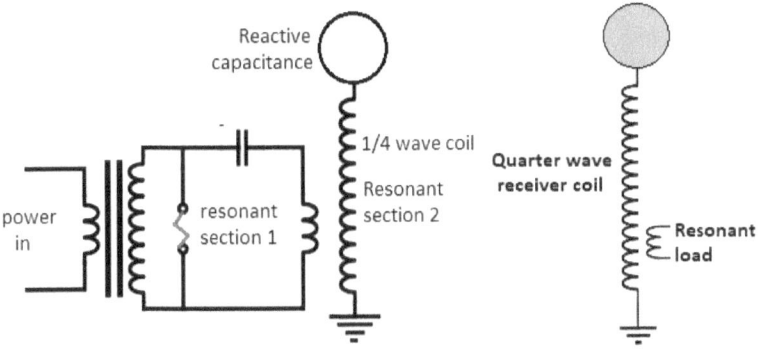

Figure 2.5 Undamped transmitter and matching quarter wave receiver

Tesla used quarter-wave antennas for resonant conduction and timing only, not for radio broadcasting. The input power signal is usually a full 360° AC signal or a square wave. Single pulsed signals work but are inefficient and run at ½ power of the 360° signals. They are not recommended for power transport. There are two power strokes, the first at 90° which charges the sphere with electrons and then the 270° stroke totally discharging it. As they say, "Timing is everything", especially in a resonating wireless power transfer system.

As I mentioned several times, wire length from terminal to the ground of the quarter wave antenna defines the wavelength (*timing*) of the oscillator. Precisely, it is the time duration for electrical [19]power to pass through its full length. Factors affecting quarter wave speed are: the conductor type, its design shape (*flat or round*), the temperature, and the insulation covering it. Power can move up to thirty percent slower than lightspeed depending on wire characteristics. To state it simply, the longer the antenna the longer the wavelength. If an antenna or conductor has no termination (*open ended*), the power wave reflects (*echoed*) back toward the source as light reflects off a mirror. Reflection is desirable in radio transmission but not with power transfer.

Radio emissions from a Reactive Power Transport (*RPT*) oscillator waste energy and upsets the FCC. They are especially undesirable in a power transfer machine. In a properly tuned Tesla transmitter, the terminal capacitor absorbs the full quarter wave signal preventing typical reflection of Hertz-style radio towers.

The main service provided by terminal capacitor is electrostatic power storage. The terminal sphere gains negative and positive charges by storing electrons or ions. In the first ninety degrees of the power input sinewave (*or square wave*), the terminal capacitor has time to store a certain amount of electrical

---

[19] Power here refers to electrical power and not electron flow. Electron flow is only a few centimeters per second.

energy. The time needed to fully charge the terminal capacitor at a given voltage is Optimum time. Optimum time is defined by several elements: the period, propagation speed, the voltage across the terminal to earth ground, linear resistance from earth to terminal, inductance, and the capacitance of the terminal.

The resonant frequency of a quarter-wave coil may be varied by several methods:

raising or lowering the terminal in reference to the earth
varying terminal capacitance (*sphere size or higher value vacuum capacitor*)
varying the induction of the quarter wave coil
applying ungrounded RF wrapping to the ¼ wave coil
reduce the parasitic capacitance of the ¼ wave coil
changing the characteristic of the ¼ wave coil (gauge, wire style, wire type, etc.)
apply cryogenic methods to lower resistance.
change the quarter wave wire length

The coiled section of the Tesla antenna is responsible for the electrical inductance. The antenna wire length determines the basic resonant frequency. A major problem in a non-cryogenic, Tesla transceiver is parasitic capacitance in the windings of the quarter wave coil. Proper coil winding techniques reduce this issue, but care and experimentation are needed. A method I devised is the five-four-two band wrapping. The winding begins at the bottom and progresses upward to the top of the coil frame. This works well on smaller diameter coils of four to six inches. Double-stick tape (*I prefer carpet tape*) is wound in sections around a PVC frame as the coil winding progresses. Wind five turns, then four turns in the grooves of the five wires, then two windings on top of those four grooves. I mean by the word groove the valley formed by two closely wound wires, or may I say, between the windings. This is one winding band of eleven turns.

The effect of parasitic capacitance in the quarter-wave coil causes an electrostatic field around the coil windings. The electrostatic field links together from the bottom to the top of the coil. Magnetism does this too, but it circles the wire and links together causing alternating magnetic poles. The electrostatic field causes power leakage in the coil outward, inward, upward, and downward depending on energy flow direction. The 'E' field from the coil presents an alternate path for power to travel. This circumvents natural resonance, self-induction, and energy flow through the wire of the coil. It also alters the resonant frequency, and consumes energy meant for transport. The parasitic capacitance increases with high dielectric insulation, wire profile, and closeness of the windings. It can be circumvented with capacitors across the windings and more is said about that later.

Linear resistance of the antenna wire should be small as possible. Use large gauge wire or flat wire to maintain quick charge/discharge time for the terminal capacitor. Effort must be made to keep the RC time constant of the antenna low. The value is calculated by T = R x C (*time = resistance times capacitance*). All windings must progress from bottom to top and should never be reverse wound. Electrostatic power moving up and down the ¼ wave coil is proportionally cancelled by each reverse winding but it also shunts power going to and from the terminal.

A large potential difference exists between the windings at the terminal and the ground connection. Electrical arcing can happen between windings due to these high potential differences (*dielectric breakdown*) ruining the coil. The wire used in most windings has a dielectric coating. The dielectric breakdown voltage of the wire must be known in order to prevent the damage.

To calculate the coils voltage per winding, divide the maximum voltage difference from ground to terminal and divide the number of windings of the coil. The dividend is the voltage across each winding. If the voltage exceeds the breakdown voltage of the wire's dielectric strength then replace the wire or reduce operating voltage.

A high L/R (*inductance divided by resistance*) factor increases system power and efficiency. The quarter wave coil of high inductance with low resistance boosts efficiency. Tesla referred to this calculation as magnifying factor or Q_m. The higher the L/R (Q_m) Factor the lower the quarter wave wire resistance. A high Q_m means higher charges on the terminal capacitor and increased efficiency. A low L/R (Q_m) suggests poor efficiency, avoid it. Tesla patented and used supercooled coils to improve the Q_m of his wireless machine, including the one at Wardenclyff; there will be more on this later.

Oscillator frequency review:

To reiterate, the frequency of the basic quarter wave antenna is a simple measurement of its length. For example, consider a spirally-wound, quarter-wave antenna of 500 feet long. The wave length (λ) of the coil is 2000 feet (*4x500*). We can choose at this point whether to use meters or miles; I will choose miles. Two thousand (*2000*) feet is .379 miles (*2000 ft/5280 ft/mi*). Now using the speed of light at 186,000 mi/sec, divide it by .379. The result is: $186000_{mi/sec}/.379_{mi}$ = **490,765 Hz**. An oscillator with a quarter wave antenna of 500 feet will have the basic frequency of 490.765 kHz without a terminal. Because the quarter wave antenna is a coil the inductance is larger and resonance is different than a long straight wire. Calculate the resonant frequency of the oscillator using antenna inductance and terminal capacitance. The base frequency is the quarter-wave length. I use an inductance meter across the quarter wave coil with the terminal disconnected to determine inductance. If inductance and frequency are known then use the resonant frequency [20]formula to calculate terminal capacitance for those values.

---

[20] http://www.1728.org/resfreq.htm

Scale the oscillator to deliver the planned power output at the chosen frequency. This can mean reducing or increasing the coil or terminal capacitance. Capacitance is the main component defining power transfer capability. You must know the capacitance (*coulomb storage in Farads*), the frequency of the oscillator, and maximum voltage of the secondary. The capacitance of each transceiver must be able to support fifty percent (*50%*) of the power to be transferred. If necessary, properly scaled inductors can be added in series with the quarter-wave coil to tune the frequency.

Parasitic capacitance can be reduced by adding capacitors across several windings progressively up the coil to the terminal. This is a way to have higher efficiency in the oscillator. The inductance of the coil must be known. Divide the total inductance by the number of windings to determine the inductance of each winding. Place capacitors at points on the winding accordingly. Each section patched should resonate at the transmitter frequency. In effect, each patched section has zero impedance, but the linear resistance is not affected.

Let's assume we wound the antenna presented (*490,765 Hz*) in the previous paragraph of 500 feet of #20 wire on a ten-inch diameter form at .007 inch between windings. Calculation shows that each winding is 31.4 inches and there are 6000 inches in 500 feet. This works out to about (6000/31.4) 191 windings. The height of the coil is about 7.5 inches from calculation, #20 wire is .032 inches in diameter plus the separation of .007 inches at 191 turns is about 7.5 inches.

Using the inductor [21]calculation for an air core coil, the inductance is ≈ 7600 μH. To resonate the oscillator at the quarter wave length (*490,765 Hz*), solve for capacitance with the resonant formula, $F_r = 1/2\pi VLC$. Calculations resolve that a 13.8 pFd (*13.8*10^{-12} Farad*) terminal is needed. If we decided to slow the resonance to 400kHz we would need a 20.8 pFd terminal. From this short drill, you can see there are calculations to do, but they are simple, algebraic ones.

Moving on:

The Tesla oscillator is a reactive power scheme. Reactive power in commercial power transmission terms is in [22]volt-ampere reactive (*VAR*). VAR is a unit by which reactive power is expressed in an AC electric power grid. The VAR in a utility power grid is ninety degrees out of phase with the actual AC supply power. The reactive power returns to the generator by a rebounding effect, every half cycle at 90°-180°, aiding the power generator.

However, in a resonant Tesla Reactive Power Transfer (*RPT*) scheme, current and voltage are 180 degrees out of phase between the transmitter and receiver. The terminal spheres always have opposite charges at any given time except at zero Volts. There is a quarter wavelength packet of electrons residing in one negatively

---

[21] https://www.daycounter.com/Calculators/Air-Core-Inductor-Calculator.phtml
[22] Wikipedia

charged sphere, and a quarter wavelength packet of positively charged ions in the other. The AC power on the primary winding at the transmitter provides **all** the power consumed by the load. The AC power supply maintains the resonant standing wave between the transmitter and receiver. There is no power gain or free power in the standard transfer scheme.

To aid in understanding the Tesla reactive power system, I have included figure 2.6 - 'A' and 'B'. The top figure 'A' is the fluid analog of the electrical system in figure 'B'. Both circuits represent functional reactive systems. In figure 'A', the piston is a solenoid driven up and down by AC power placed on the input coil. The rubber bladders on the analog diagram represent the terminal capacitors on the electrical version in diagram 'B'. When the solenoid piston is driven up or down, the compressed fluid fills or exits the bladders in an alternating fashion.

The transmitter's piston drives down on the first stroke expending power to expand the receiver's bladder. On the sequential stroke, the power invested in expanding the bladder is returned to the circuit, assisting the piston and saving energy when it re-inflates the transmitter's bladder. In simpler words, the power required to drive the next half stroke of the piston is minimal once either one of the bladders are charged with fluid. This is the basis of the reactive power transfer system (*RPT*).

The capacitive terminals of the RPT serve the same purpose as the bladders on the analog fluid circuit. The electrons that fill the terminal *of the receiver* are compressed with voltage, and when the voltage reverses, the *receiver's* terminal capacitor discharges, aiding the transmitter primary power to refill the transmitter terminal capacitor.

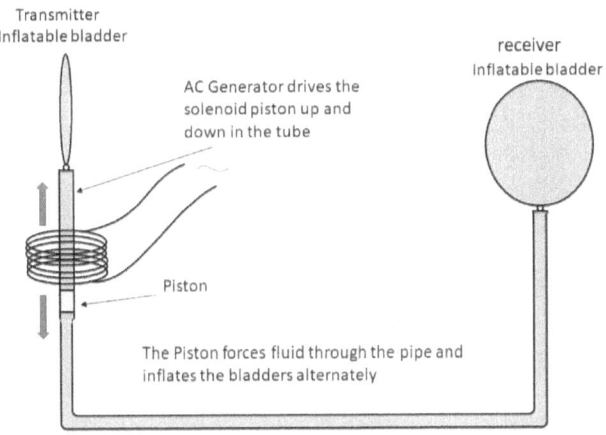

Figure 2.6-A

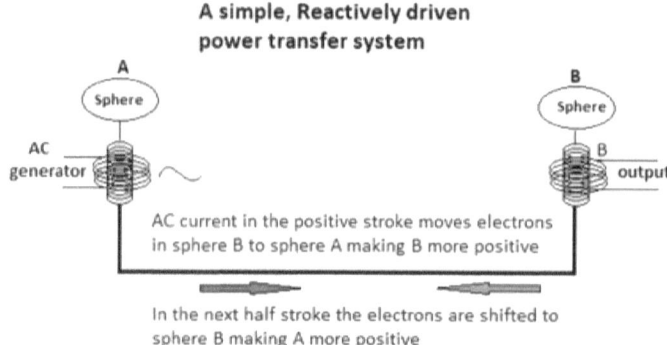

A simple, Reactively driven power transfer system

AC current in the positive stroke moves electrons in sphere B to sphere A making B more positive

In the next half stroke the electrons are shifted to sphere B making A more positive

Figure 2.6-B Is the electrical equivalent of 2.6-A

Analogies can be stretched only so far. This is an effective example of the transfer process in a Reactive Power Transport circuit (*RPT*). The enormous, explosive power of moving electrons under great voltage, and the standing wave's constructive, power charging cannot be explained by an analog. The longitudinal electron wave produces a very small magnetic field, if any, as it moves across the NM (*Natural Medium*).

The right-hand rule of magnetic fields moving in a conductor show the direction of the magnetic field. If this rule is applied to the longitudinal electron wave (LEW) the magnetic fields are canceled in the medium within the individual concentric compression rings. The field folds upon itself within the compression wave.

The **'output'** (B) on the right in figure 2.6-B is passed to the load by transformer action of the magnetic field in the quarter wave coil of the receiver. To increase the efficiency of the output, the load is also resonantly tuned to the operating frequency.

Chapter 3: Introduction to the Natural Media (*NM*) and Reactive Resonant Systems

The Natural Medium (*abbreviated as NM*) was the name Tesla used to describe the earth with its electrical properties. In the late 1800s earth conductivity was already used by telephone, telegraph, and power companies as a return conductor for their systems and for grounding protection against lightning. Figure 3.1-A illustrates the path of returning power via the earth when used as a second conductor by a power company. Telephone and telegraph companies still use this scheme today.

Single wire earth return system

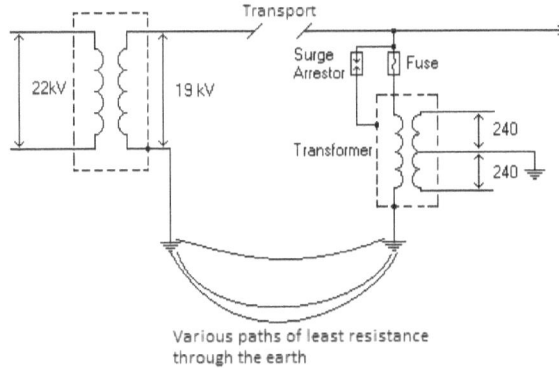

Various paths of least resistance
through the earth

**Figure 3.1-A Earth as a return wire,
used by some power companies**

The telephone companies' standard POTs (*Plain Old Telephone service*) telephones connect to the central office by copper cable. Telephone companies use forty-eight-volt direct current for voice transmission, one hundred-five volts at twenty hertz for ringing, and a ground return is used for forty-eight-volt coin collection in the old payphones. Even at these low voltages, a telephone ringer for a party line can ring across one wire to earth ground. A small signal current measured in milliamps returns to the central office through the earth some miles away with no measurable loss of energy. The earth offers little resistance to low frequency current flow, about .005 ohms from point to [23]antipode. Tesla was a telephone man in his early career and he was keenly aware of the earth's conductivity.

Tesla's wireless engineering plan required certain knowledge about the NM that was unknown at the time. His questions were:

---

[23] Antipode is the side opposite on the earth

1. What AC frequencies could be efficiently passed from point-to-point through the medium
2. How far would the signal pulses travel at a given voltage
3. What were the of the electrical properties of the medium (capacitance, inductance, and resistance).

Engineering demands precision values and these were still unknown. Knowledge of earth constants and variables were essentials to engineer the best machine for wireless power transfer.

Earth return power depends on a robust ground; it is of utmost importance, and difficult to obtain in some places. In the western U.S. there are nonconductive areas and rocky mountainous places. But, in the central and eastern US a good ground connection is easily obtained. Tesla Reactive Power Transfer (*RPT*) oscillators must have firm earth grounds for efficiency and standing wave creation. Poor grounding results in heating and power loss at the ground rods or plates. Tesla knew this and about the importance of proper 'earthing'. This is written about in the Colorado Springs Notes journal as a major challenge.

At his Wardenclyff power station on Long Island, NY, he spent most of his grant money from Morgan on the grounding scheme for the station. He dug down 120 feet into granite, and excavated radial tunnels connecting cables and iron ground rods to his Transmitter. He also provided ventilation tunnels upward, angling toward the surface. They terminated at the surface in rounded top structures with window-like openings. This design, like a prairie dog tunnel, created a slight vacuum from the passing wind to evacuate the heat and steam produced by the high-current ground boundary of the system. The ground got very hot from the three to twenty million-volt power currents the tower produced. The heat caused steam from the water table to form in the tunnels and under the tower. Though Tesla never officially opened the tower for production, he actively tested with it. In an interview he said it cost a hundred dollars per day to run the structure for testing; this is in the early 1900s when the average annual income was only $450.

The NM capacitance (*earth capacitance*) is 710 micro-farads as calculated by Tesla and various modern scientist. The terrestrial capacitor's negative plate is the earth itself. The positive plate is the conductive upper atmosphere (*Ionosphere*) and it maintains a positive charge by cosmic particles and the solar wind bombardment.

When a charge is placed on one plate of a capacitor the other plate is charged with an equal and opposite charge. The (+) positive Ionosphere draws interstitial electrons from the earth's inner regions to the surface creating a negative charge of equal and opposite potential on the surface. These electrons are the medium of the Longitudinal Electron Wave (*LEW*).

The earth's charge is retained by the nonconductive air dielectric, see figure 3.2. [24]Estimates of the electrical energy stored in the Earth's dielectric atmosphere is around [25]150 gigajoules (*GJ*) with a net negative electric charge of 500,000 coulombs over the earth's surface.

Spherical capacitor and electrical equivalent

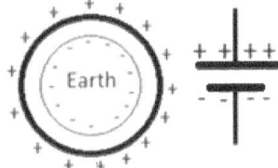

Figure 3.2 Earth capacitance

In Figure 3.2 to the right is the electrical equivalent of the spherical capacitor. A signal placed on the negative or positive plate of a capacitor electrostatically impresses it onto the other at 180° out of phase. This applies to the earth. The earth is a conductor with self-inductance and a resistance of only .005 Ohms from any point to its antipode. The earth's inductance is 0.586 Henrys calculated using the Schumann frequency of 7.8 Hz and a capacitance of 710 microfarads. Tesla calculated the [26]earth's capacitance ($C = 4 \pi \varepsilon_0 * R$) and its inductance, and he used these values to calculate the resonant frequency of the earth many years ahead of Shuman.

Tesla used this resonant frequency to exactly measure the circumference of the earth. The measurement varies according to the origin of the measurement, slight bulges exist in places. To calculate the earth's resonant frequency, use the formula: $\lambda = c/C_e$, Where, λ = wave length, c = the speed of light in miles per second, C_e is the circumference of the earth in miles. Substituting values, the calculation returns the resonant, first harmonic frequency (*to the antipode*) of the earth; 186000 / 24901 ≈ 7.4695795349584 Hz. The second harmonic of 14.9 Hz creates a full standing wave between a transmitter and its antipode.

The earth's total reactance (X_t) is about 55 Ohms at the second harmonic of 15 Hertz. At **500,000 Hz** the inductive reactance X_L is about 1,840,000 Ω (*Ohms*) ($X_L = 2\pi fL$). The capacitive reactance (X_C) ($X_C = 1/2\pi fC$) at the same frequency is less than 0.005 ohms. This indicates that high frequency capacitive coupling could be possible over great distances. Tesla decided to use low frequencies and the earth conductor rather than the high frequencies and the air insulator for power transfer. He never said why.

---

[24] The Earth's Electrical Surface Potential A summary of present understanding (January 2007) by Gaetan Chavalier, PhD, Director of Research, California Institute for Human Science, Graduate School & Research Center, Encinitas, CA

[25] https://en.wikiversity.org/wiki/Natural_electric_field_of_the_Earth

[26] http://hyperphysics.phy-astr.gsu.edu/hbase/electric/capsph.html

Capacitive coupling is possible, and it is a basic part of Tesla's wireless power plan for aerial vehicles and automobiles examined in chapter five. Capacitively generated longitudinal waves in the dielectric (*air*) are called [27] Zenneck surface waves. They take advantage of the low X_C of the earth at high frequency. They are transmitted in the 200 kHz to low MHz range and use the earth's conductive surface as a waveguide. Zenneck waves do not have the great range of the inductive longitudinal compression wave, but they can deliver power effectively with a proper receiver.

High frequencies, above 800 kHz were not a part of the Tesla plan because the high inductive reactance of the earth at that frequency would not permit effective transport of the longitudinal wave, known as the LEW (*Longitudinal Electron Wave*). The LEW is the transport vehicle for wireless power, and will be discussed in detail in chapter four.

Showed in figure 3.3 is a different example of the capacitive equivalent of the earth. The surface has a negative charge and the Ionosphere a positive charge. The air/atmosphere is the dielectric. I mentioned earlier that a signal impressed to the earth plate moves over the whole earth surface because of its conductive properties. Capacitive coupling transfers that signal to the Ionospheric plate at 180° out of phase. The signal must have proper scale to overpower electrical noise because of the size of earth. Any conductor suspended in the dielectric (*atmosphere*) is excited by the varying charges. We exploit this force to provide power to aerial apparatuses.

Earth as a Capacitor

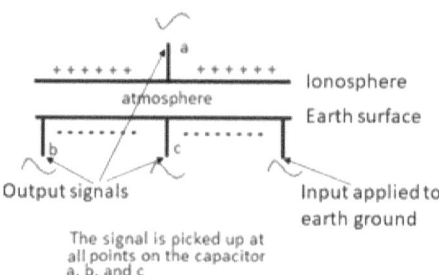

The signal is picked up at
all points on the capacitor
a, b, and c

Figure 3.3

Our test showed the Tesla resonant Reactive Power Transfer (RPT) will not efficiently transfer frequencies of one MHz and higher without significant voltage (*20kV and up*). We did tests to determine if power is capacitively transferred between spherical terminals. We tried various frequencies from 2000 Hz to 1.5 MHz. Tests indicate that there is no capacitive coupling between the Tesla RPT system's spherical capacitors unless they are within a quarter wavelength. Most of this electric field coupling was from parasitic capacitance in the quarter wave coils and not the

---

[27] https://en.wikipedia.org/wiki/Zenneck_wave

terminal spheres.

The power transfer between RPT transceivers moves through the earth in longitudinal waves, also known as compression waves. We did not experiment with Zenneck waves except at short range and they were effective at transporting power only on a grounded system. We did not use a high frequency signal in the high megahertz range required to launch a Zenneck wave with our apparatus.

The Tesla RPT transmitter and receiver *cannot* share ground rods or be in the same grounding plane (*conducting ground space*). The earth connections between oscillators must be separated a minimum of 1/8 wavelength. Within the 1/8 wavelength range both transceivers are likely within the same node (*zero-volt*) grounding field; my term for this is **node concurrency**. Two transmitters can share the same ground, and two receivers can share the same ground, however, resonant transmitters and receivers operate 180° out of phase. Resonance between oscillators is not possible in the node-concurrent design or in other words on the same ground rod or in the same [28]grounding field.

If local (*within 1/8 wavelength*) testing is necessary, remove ground connections on each oscillator and connect them with wire. Protect the wire from all earth ground potential. The wire will provide the channel for resonant testing. Bear in mind that magnetic and electric field induction between the units can, and probably will happen. Testing two oscillators within 1/8 wavelength separation with frequencies above 350 kHz raises the likelihood of electric field coupling. Improper separation of the transceivers causes faulty and meaningless test results. The only exception is purposeful 'Resonant field' coupling for short-range power transfer. A wireless static-electric charger has a range of three hundred feet or more when connected via a resonant reactive receiver.

We examined the RPT units for emissions of RF. The small section of wire between the middle of the quarter wave winding and the wire connecting to the spherical capacitor emits some radio interference. The radio emission was low power and twice the resonant frequency of the oscillator. RF interference can be reduced with ungrounded metal foil shielding at the top of the coil. The shielding will raise self-resonance of the quarter wave coil by up to three times depending on wrapping technique. However, the resonant frequency of the RPT transceivers will not change.

Our tests proved without question earth ground is the only medium involved in transferring resonant power. The exception is the electron plasma waves (EP) which will be the focus in chapter five. There is no capacitive power coupling between the spherical terminals of the transmitter and receiver unless they are close.

Our prototype RPT oscillators used a 10"x12" copper ground plates in a saltwater marsh in combination with ten-foot earth rods at both ends. Various

---

[28] Grounding field is an area near a ground rod. The radius of the field encircles the rod and any rod driven within the field is considered to be the same ground point.
http://www.weschler.com/_upload/sitepdfs/techref/gettingdowntoearth.pdf

arrangements of this scheme worked well. We found that a copperplate on the transmitter in saltwater, and a ground rod in the earth at the receiver end provided the highest voltage levels on the receiver output. The copperplates at our labs were in the tidal waters of the Lafayette river (*a saltwater river*) in the Norfolk, Virginia.

Test prove a short, poorly grounded rod increases the voltage on a receiver, but there is less power available because the voltage drops with a small load. We found this an advantage for transceivers used in communication-only channels with resonant or capacitive coupling. However, I recommend the best ground possible for all types of systems except aerial. Signal voltage is boosted easily with a proper transformer design.

Tests at various frequencies were tried using *closely coupled* coils wound on an air core, PVC, frame. This is a classical arrangement and quickly modifiable compared to ones we wound on a ferrite core. We built the resonant receivers and transmitters with the same electrical characteristics. They were operated with various multi-frequency power supplies at the transmitter. The types used were: an HP 214b (+/- *pulse generator*), a Global Specialties 4005 arbitrary waveform generator, and several custom-made variable frequency MOSFET, square-wave pulsed power supplies. We placed a transmitter and receiver at 500 feet apart connected to both ground rods and water plates. A 1 MHz signal at 10 kV was sent from the transmitter. At the receiver no signal was detected but at 500 kHz the signal was fine and we received a higher voltage than the supply at the transmitter. This was an effect of resonance.

We swept through various frequencies. The receiver voltage rolled off at about 650 kHz. We sustained resonance by lowering or raising the spherical capacitance above the earth with a pulley, changing the terminal capacitor, and using adjustable length inductors in the ¼ wave antenna.

The HP214b is a high-voltage pulsing machine and cannot make sine waves. The pulses and power are adjustable for pulsed square waves and sawtooth waves. We tried various waveforms and found that half-square (*+ or -*) wave pulses delivered more power. The duration and pulse width are adjustable on the HP214b. We sent short bursts and long bursts. Longer burst worked better up to a certain point then the signal dropped out at the receiver. Skewed sawtooth patterns were low or no power, but strait sawtooth worked well and was the closest we could get to a sine wave from the machine. Adjustment of power level and signal form at the transmitter end instantaneously displayed on the oscilloscope at the receiver fine tuning easy.

As a side note: in chapter seven I provide a method to send LEWs to a resonant receiver without using a resonant transmitter. This is a great labor-saving method for adjusting receiver resonant frequency while connect in an operational mode. It a precision and easy method for short ranges. The method also proves the existence of longitudinal waves and their power delivery.

We experimented with moving the primary winding of the transmitter at various positions up and down on the quarter-wave coil to see how output was affected at the receiver. Power output increased as the primary was moved upward toward the center. After passing center, power dropped at the receiver. The difference was three to four volts. This seemed an insignificant amount to call for an oscillator design change.

Capacitive spheres are an essential part of the RPT arrangement and their capacitance must match the power capacity of the oscillator at the working frequency. Vacuum tube capacitors (*single plate*) can replace spheres in most applications but are almost impossible to find online. Tesla did extensive testing on the terminal sphere in Colorado Springs. He found the terminal capacitance increases by 2.7 percent per foot raised above the earth. We confirmed Tesla's findings but our capacitance increased about 2 percent each foot. We tried discarding the terminal capacitor in the oscillator; it cut power yield and caused radio interference at the transmitter and arcing at the terminal point.

The electrical current (*coulomb*) requirement for the sphere is calculated on a fifty percent duty cycle (*or, the power delivered in one-half wave*). This means the terminal sphere must be able to absorb one half of the current (*coulombs*) of the resonant oscillator per cycle. The sum of the voltage between the terminal capacitors of the two resonant oscillators is the full wave, peak-to-peak (*360°*) voltage of the secondary: $V_T = |V_t| + |V_r|$: where V_t and V_r are transmitter and receiver voltages. This is the total working voltage in one complete LEW cycle of 360°. The total, peak-to-peak voltage (V_T) specifies the electron displacement (*the maximum to minimum electron movement*) within the LEW compression wave. Resonant reflection (*ERL feedback*) creates the standing LEW. The LEW's voltage amplitude will quickly increase by many times as the result of ongoing [29]constructive interference.

An RPT oscillator should be oversized by twenty percent of the target output. Greater current, demands a larger terminal capacitor and higher voltages at the transceivers (*a pair of oscillators resonantly connected*). A terminal capacitance which is too large and/or a voltage too low makes the power signal [30]'Q' drop. This is an effect of improper scaling. The low 'Q' quashes the advantages of electrostatic elasticity in the terminal. The same effect is simulated by enlarging the rubber bladder of figure 2.6-A (*in the previous chapter*) and not increasing the piston's fluid capacity. Optimization of the Terminal capacitor is paramount.

High concentration of electrons, or ionic (+) charges is desirable because the higher the Q the more reactive current moves between the transmitter and receiver increasing the power and transport range. Higher voltage between the oscillators' spheres increases transfer efficiency. The primary power supply at the transmitter combines with the power in the standing wave raising voltage on the terminals. A large power transfer ability for the Tesla RPT requires a higher driving voltage on the transmitter to achieve it. Terminal capacitance is key in defining coulomb/power capacity (*power - I^2R*) of the oscillators.

---

[29] http://www.phys.uconn.edu/~gibson/Notes/Section5_2/Sec5_2.htm
[30] https://en.wikipedia.org/wiki/Q_factor

Terminal capacitor elasticity has two parts, the voltage potential and the dielectric covering of the terminal capacitor. The voltage impressed on the sphere creates a charge (*Q*), and the sphere insulation (*dielectric*) is the reinforcement for containing the charge and preventing breakout. When high-pressure (*voltage*) compresses more electrons onto the sphere than the insulation can physically contain, dielectric breakdown occurs, and sparks spring from the sphere upward, or from earth to the sphere. The breakout situation can shock or kill people near it. Breakout triggers an RSC (*resonant signal collapse*) event dropping the standing wave and crashing the system. Vacuum tube terminals can prevent breakout.

Listed here are some important points to consider about RPT terminal capacitors:

1. electron Q (*coulomb*) density of the sphere increases with a higher voltage on the secondary
2. electrical arcing must be prevented
3. the terminal's electron supply comes from a well-grounded earth connection
4. the secondary quarter wave wire must have a low resistance from earth to the terminal

I stress the point, the single-plate, spherical capacitance (or *vacuum tube*) of the Tesla resonant RPT structure defines the coulombs of power at the receiver, and the power (I^2R) in the standing LEW. The higher the resonant frequency, the smaller the terminal capacitance required to transport the same power equivalent but transport range is sacrificed. All the component parts must work in harmony to form a dependable and efficient reactive standing wave power transport assembly. Tesla said, I will paraphrase, "each section of the wireless system must be resonant at the same frequency or an odd multiple of that frequency".

The four resonantly tuned RPT system sections are:

> The power input section and the primary winding
> The secondary of the transmitter coil's inductance and the terminal capacitance
> The receiver primary coil inductance and terminal capacitance
> The receiver's secondary coil inductance and the driven load at the output

Operational frequency and power needed at the receiver defines the size of the terminal capacitor. As mentioned earlier, in an RPT design, plan for about twenty percent more charge density in the terminal than needed. Equation 1 of [31]Maxwell's (*gauss' law*) equations [$\int_v (\nabla \cdot D) \, dV = \int_v p dV$] or simply $\nabla \cdot D = pV$ is used to calculate the volume charge density of the terminal capacitor (*vacuum tube or sphere*). The equation says a larger sphere can contain more charge. The divergence (*D*) increases with voltage (*E - EMF*): 1V = 1 joule/coulomb. The equation

---

[31] http://www.maxwells-equations.com/

E=W/C is derived from the power formula P=IE and it defines the terminal voltage: where E is EMF, W is in Watts and C is amperes (*Coulombs*).

The voltage necessary to add the desired charge per cm^2 on the sphere can be calculated. The maximum dielectric breakdown voltage of the terminal capacitor covering is needed. There are charts available online with the [32]dielectric value of various materials to do calculations. Do not exceed the breakdown voltage of the dielectric or the electrostatic charge will break out of the sphere like a bolt of lightning. Reiterating for safety's sake, this power leak can kill or cause bodily injury and cause loss of connectivity to the receiver.

The permittivity of the sphere's dielectric covering is constant. Its physical condition, thickness, polarization density, or operating and breakdown voltage can change over time. Outdoor systems become undependable after prolonged weather exposure. Dielectric breakdown from environmental surface weakening, closeness to a grounded surface, or other physical damage happens. Transmitters and receivers should be inspected regularly. Ours had peeling of the titanium-dioxide covering from sun exposure after a year.

Powering-up the transmitter's primary transformer starts the Tesla RPT system. Within one second the transmitter and the receiver terminal and coils are fully energized. Power in the standing-wave LEW is the sum of the power in each oscillator. The oscillator's power output is a combination of the discharging terminal capacitor, the collapsing secondary magnetic field, and power from the primary power supply. These sources work together to cause a shock wave of power into or out of the earth. The combination of this power creates and sustains a resonant, standing wave between the transmitter and receiver. The electrostatic and magnetic fields oscillate between static and active power supporting the reactive transport link. The transmitter's primary power source provides all the consumable energy at the receiver. Overdrawing the power will cause a condition of RSC (*resonant signal collapse*) and drop the load. Rebooting the transmitter is usually required after correcting the cause.

The terminal capacitor – a closer examination:

Defining terminal capacitor size requires understanding of the charges to be put on it. One electron has a charge of $1.602*10^{-19}$ coulombs (*C*). One ampere of current is $6.242*10^{18}$ electrons per second crossing a point in a conductor. One Q (*or coulomb*) charge on the terminal capacitor is $6.242*10^{18}$ electrons, or one Joule of power. One Farad (*Fd*) is 1 coulomb of charge of electrons per volt placed on a surface in 1 second.

Consider a terminal sphere with a diameter of twenty inches. The spherical capacitance is $C = 4\pi\varepsilon_0 R$: substituting values $4\pi(8.854187817*10^{-12})* 10 \approx 9619 * 10^{-12}$ Fd (*9619 pFd*) not considering height above the earth.

---

[32] https://www.kabusa.com/Dilectric-Constants.pdf

A 9619 pFd (.000000009619 Fd) single plate capacitor can contain a charge of .000000009619 coulombs ([33]Q) at 1 volt in one second: at just 20,000 volts it becomes 0.000000009619 * 20000 ≈ .000192 Q. This figure represents 20,000 volts per one second of DC charging time, however, we are using AC, not DC on our system. A frequency of 250 kHz (*randomly selected*) is a common frequency for a low voltage (*20 kV*) RPT arrangement with about a twelve-mile range.

At 250 kHz the AC current changes direction two times per cycle, positive to negative. The terminal capacitor will charge and discharge 500,000 times per second (*250,000 * 2 = 500,000 oscillations*). The RC time constant can inhibit a full charge, so be mindful of linear resistance in the coil. The total current capability of the system is 500,000 x .000192 *coulomb* = 96 Amps of current per second; this is optimum with proper conductor scaling. The LEW voltage is 40,000 volts or more; the 20-kV secondary voltage is at least doubled by resonance in the standing wave. Twenty-thousand volts on the quarter-wave coil is the value I will use in this example.

LEW [34]Power Displacement Calculation: *These figures assume the following:* *the power supply is adequate; the quarter wave windings and terminal can support the current flow and operating voltage used in these examples.*

C = 0.000000009619 per terminal therefore C = 1.9*10⁻⁸ Fd- *For **Two** Transceiver Terminals*

F = 250,000 - *Frequency of the LEW*

R = 5.005Ω - *Linear resistance of the two ¼ wave coils (5 Ohm + earth resistance .005 Ohm)*

I = 96A – *current in the transmitter per second at 250,000 Hz*

V = 20,000 V – *voltage on the secondary, not V in the LEW*

$X_C = 1/2\pi FC = 0.0007654557$, *Capacitive reactance*

$X_L = 2\pi FL = 6.28 * 250000 * 0.586 = 920486$, *Inductive reactance of earth*

X = 920486 - 0.0007654557 = **920,485** Ω *Total reactance*

Z_t = 920,485 *total impedance: coil resistance plus earth resistance.*

The [35]VAR formula is (**Q = I²X or E²/X**) then **Q** = E²/X = 20000² / 920485 = **434.55**

Q = 434.55 VAR

The standing wave gain was not used in the calculation, it is normally doubled in a resonant LEW. This is not usable power. It provides the power for the transport link through the natural medium. The Apparent power of the system is also important in AC power transfer.

---

[33] C represents capacitance and Coulombs. One Coulomb of charge is labeled Q.
[34] Jtotal = Jconduction + ∂D/∂t   (Total Electric Current) from
https://www.researchgate.net/publication/302966559_Maxwell's_Original_Equations
[35] https://en.wikipedia.org/wiki/AC_power

Formula: Apparent power = $(I_Z)^2 * Z_t$ (from $P = I^2R$)

Where $I^2 = 96^2 = $ **9216**

Z = total impedance = $\sqrt{(R^2+X^2)} \approx$ **920,485 Ohms**

The **Apparent power** (S) is $(96)^2 * Z = 9216 * 920485 = $ **8,483,189,760 VA**

Current = 96 Amps per cycle (@250,000 Hz in each terminal capacitor)

True power $P_T = I^2 * R = (96)^2 * 5.005$: P = **46,080 Watts** in the **LEW**

The energy density increases through the Tesla scheme by frequency multiplication. It is defined by the total amperes, charge plus discharge, of the terminal capacitors per second. [36]True power is consumed from the transmitter's primary winding through the LEW at the receiver's secondary winding. A power inverter at the receiver supplies DC or sixty cycle power for consumer use. Clearly you see why resonance is so important across the load and primary windings. Resonance preserves a strong standing LEW and lowers losses at the resonant secondary by maintaining constant phase and increasing load impedance (*tuned power*).

The available power for continual consumption is the power applied at the primary winding of the transmitter. The power in the LEW is usable, but it can drain quickly. To maintain the LEW and reactive power transport, the best practice is to never draw more power out than the **True power**. This is not a 'free power' arrangement. The LEW's standing wave power-doubling is a bonus because voltage build-up maintains the resonant LEW and provides 'free' transport. The higher voltage enhances the Electron Plasma field used to power aerial systems; this is the real bonus of the LEW.

The next time you are asked how did Tesla get more power out than he put in, show them this mathematical example of True power. The resonant arrangement multiplies current density in transport by constructive interference. Usable power for the consumer must be provided in full at the transmitter. At low frequencies like sixty Hertz there would be little gain because the wavelength is so long (*3,100 mi.*). This made Tesla excited about his high frequency, reactive power transfer. The unexplainable power increase is from the endless supply of earth's electrons to the resonant LEW and the high frequency.

In a normal transformer coil the number of electrons moving in the circuit is limited to the mass of the copper in the coils. In the Tesla Reactive Power Transfer system, the conductivity of the quarter wave coil and the storage ability of the terminal capacitor are the limiting factors, electrons from the earth are abundant. Ohm's law and Kirchhokoff's law combined with the power from [37]resonance explains the abundant power in a concatenated, resonant RPT (*Reactive Power*

---

[36] http://www.tpub.com/neets/book2/4g.htm
[37] https://en.wikipedia.org/wiki/Resonance

Transport) scheme. The currents are produced thousands of times per second multiplying the power.

Figure 3.4 shows a pair of 40 kW (*calculated by terminal capacitance*) resonant RPT oscillators about three feet tall. These are prototypes of an underwater light speed communication and low power transport trial for a university in Boston and similar in size to the example just calculated. The arrangement functions on low secondary voltage of 10-kV on the quarter-wave coil. The ground plate is a ten square inch copper sheet immersed in seawater. The quarter wave coils are made of #20-gauge plastic insulated wire and cannot support more than 500 volts between windings.

Figure 3.4 Forty-kilowatt transmitter and receiver

The power of the AC currents entering the earth or water does not kill animals, they are non-lethal LEWs (*Longitudinal Electron Waves*) in the earth. LEWs are all around us, but we are unaware of them because they are frequency multiplexed on the earth ground. LEWs are created in the NM similarly to the way a TDR (*Time Domain Reflectometry*) signal is created on a cable to test its length.

LEWs produce nonthermal plasma waves on the earth's surface. In the next two chapters, the Longitudinal Electron Wave (*LEW*) and the electron plasma (*EP*) field will be exposed in detail. The earth LEWs are not a new phenomenon. They are rarely mentioned in physics, and very few electrical engineers even know about them. Physicist studying lightning and electrical properties of the atmosphere aware of the LEW as an electrostatic charge. Its power, and various wavelengths have been measured. A search of the longitudinal waves usually points to borderline or para-science web sites *(not a bad thing)*. A close look at the underlying physics that describes the properties of the LEW and its behavior is coming up next in chapter four.

Chapter 4: The Longitudinal Electron Wave (LEW)
The means of power transport through the Natural Medium

What seems impossible is possible. I learned earth ground was a sink for electrons and a return conductor, well it is that, but much more. At Bell Communication Research and Telcordia Technologies I served as director of R&D for integrated testing for voice and data communication lines. Ground potential was always part of the equation for every circuit we tested. Ground potential can ruin a voice or data link. I discovered one incredibly good thing about earth ground, it can transport electrical power like a superconductor.

In the radio, electronics, and telecommunication industry, lightning is the nemesis of dependable service and warranties on equipment. Telephone companies lose millions of dollars in equipment every year due to lightning strikes, chiefly with microwave and cell towers. The problem with protective grounding for any electrical circuitry is the LEW (*Longitudinal Electron Wave*) associated with lightning strikes.

The LEW can change an earth ground from zero potential to thousands of volts in a few nanoseconds. Resonant circuits or electrical equipment with a ground potential different from the LEW will burn out or have damage. Protection from a LEW calls for reactive surge protection. Tesla discovered the LEW and its power in [38]Colorado Springs on July 4, 1899. The full account is in the <u>Colorado Springs Notes</u>. He did not call it a LEW, but referred to it in his reference to lightning and the standing waves it created. The name longitudinal electron wave or LEW, describes the wave and its medium.

Figure 4.0 Compression wave on the surface of a pond

The LEW resembles a sound or water wave as shown in figure 4.0, but the medium is the ocean of electrons on the earth's surface. The Tesla reactive power transfer (*RPT*) system either injects or absorbs electrons at the grounding plate. This

---

[38] Colorado Springs Notes 1899-1900 page 61.

action creates disturbances in the form of longitudinal waves in the medium. The LEW travels at light speed (*or at scalar wave velocity of πc/2 around the spherical medium according to Tesla*) away from or toward the transmitter ground plate depending on the polarity of the plate's voltage. A LEW is created in the NM when an active power source is connected to an earth ground. Things making LEWs are lightning strikes, downed power lines, earth fault lines, radio towers, or a Tesla RPT oscillator.

LEWs transport power efficiently through the earth due to the high [39]Bulk Modulus of the electron medium. The LEW is either a compression or a rarefaction wave. The compression wave moves away from the ground plate with a negative pulse and rarefaction waves move toward it with a positive the pulse. I will assume the LEW travels at lightspeed (*c*): there will be more said about this later in the chapter. A transmitter's LEW compression wave moves toward its [40]antipode.

"What exactly is a longitudinal wave?", you may ask. The [41]Wikipedia definition of a longitudinal wave: *"Longitudinal waves are waves in which the displacement of the medium is in the same direction as, or the opposite direction to, the direction of propagation of the wave. Mechanical longitudinal waves are also called compressional or compression waves, because they exhibit compression and **rarefaction** when traveling through a medium"*.

To compare the power delivery of a longitudinal wave to that of a transverse wave, imagine two people holding a broom handle. The power instantly transmits at high efficiency from both ends of the handle with a push or pull. But the transverse wave's power delivery is dreadfully weak. The analogy often used in textbooks to simulate a transverse radio wave is that of a 'jump rope' or 'string' example. Waves are formed with a flip of the hand launching a wave toward the opposite transporting little power. Longitudinal Electron Waves (LEWs), however, are more solid and efficient than any mechanical analogue we may use to describe them.

The earth has a vast ocean of electrons moving in continuous undulations like the wind driven waves of the sea. Compression waves in the Atlantic Ocean can be made by a meteor splashing down or by tossing in a pebble. The former is detected over great distances, the other over a very small range before its wave signal is lost in the noise. This same analogy holds true for electric disturbances through the earth. Consider the vastness of the NM (*earth*) while we examine Longitudinal Electron Wave.

Scaling of Tesla Reactive Power Transfer resonant oscillators depends on the project. Small and short ranged oscillators are for rural zones, farms, electrifying fences, powering drones, and powering water pumps or machinery. A global-scale like Tesla's Colorado Springs, CO or the Wardenclyff, Long Island plant is for world-

---

[39] https://en.wikipedia.org/wiki/Bulk_modulus
[40] Antipode in the Glossary
[41] https://en.wikipedia.org/wiki/Longitudinal_wave

wide applications. Finally, the zones between are for medium to large oscillators used for, underwater power, drones (*subsea or aerial*), houses, businesses, broadband/Internet communication, and to charge moving cars.

In this chapter LEWs (*Longitudinal Electron Waves*) and their features are front and center. The term LEW (*longitudinal electron wave*) defines the type of electrical transmission through a medium of electrons. LEWs are mechanical moving like sound waves with rarefaction points and compression points, but unlike air molecules electrons create electric fields (*electron plasma or EP*), and slight magnetic fields when they move.

The example of a thick copper bar in figure 4.1 is a conductor with high frequency AC power applied at the ends through a capacitor. The thick copper bar shows two of the characteristics of a LEW, the compression and rarefaction waves. These waves have a shorter period and behave differently than sixty or fifty-five Hertz current.

In the example the darker bands on the bar represent antinodes (*compression*) with a high electron density and a negative charge. The lighter bands are nodes (*positive*) and characterize low electron density. There are two other familiar attributes not shown, the electric field at right angles to the copper bar, and magnetic field circling the bar in a spiral. LEWs traveling in the NM have the same basic properties as high frequency AC current flowing in a thick bar or a wire. The higher the frequency, the closer to the surface the LEW travels.

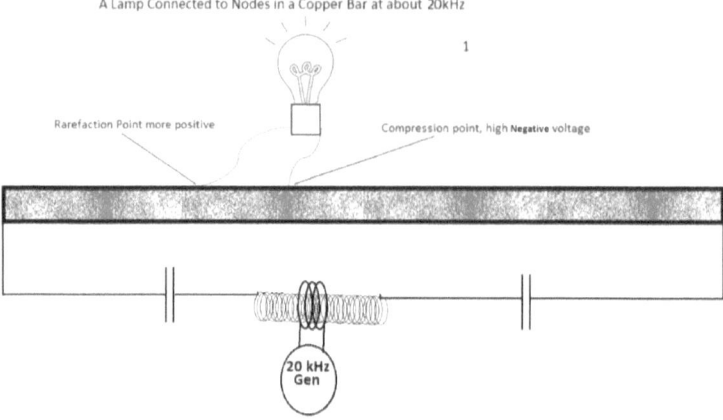

A Lamp Connected to Nodes in a Copper Bar at about 20kHz

1

Rarefaction Point more positive

Compression point, high Negative voltage

20 kHz Gen

Figure 4.1 Longitudinal wave nodes and antinodes unveiled in a solid bar conductor

As previously mentioned, there are many frequencies of naturally occurring LEWs. Lightning, volcanos, hot springs, and earth tectonic compression create them. Non-natural LEWs come from electrical power arcing to ground, AC ground return circuits, commercial power, radio and TV broadcast towers, ELF transmitters, and Tesla RPT (*reactive power transfer*) towers.

Standing wave LEWs formed by a resonant Tesla reactive power transport (*RPT*) systems produce two negative pulses per cycle. Tesla 'broadcast only' systems are low power and produce LEWs with one negative (*compression stroke*) and one positive (*rarefication stroke*) pulse per cycle. Lightning strikes usually have enough voltage to reach the antipode (*open end conductor*) and echo back; like the Tesla system it becomes a half-wave LEW.

LEWs come in two styles, the half-wave and the full-wave. Simply speaking, full wave LEWs do not resonate because the transmitter does not disturb the medium with enough power to reach the antipode or a receiver. Half-wave LEWs are more powerful because they reach the antipode or a resonant receiver and reflect creating a constructive, standing wave.

Global standing LEWs are created by high-voltages, 500,000 to ten million volts. They travel to the antipode and reflect. Reflection creates a standing wave LEW in the NM (*natural medium*) between the transmitter and its antipode. A standing wave is created by [42]superposition when two waves of the same frequency meet coming from opposite directions. The power in a standing wave equals the sum of the two waves and they continue to magnify by resonance. The Tesla RPT method exploits this effect and preserves a resonant standing wave to transport power. Other standing wave transceivers that are not global may use voltages from 20,000 to 2 million. The standing waves are established between the transceivers and not their antipodes.

Standing waves can have short durations. A good example is the standing wave LEWs created when lightning strikes. The wavelength of a lightning LEW is four times the length of the lightning bolt (the *length of the bolt is the quarter wave antenna*). The LEW will reach its antipode and resonate through the lightning bolt until the power exhausts in the terminal cloud. The standing wave oscillates from the cloud to the antipode through the lightning bolt. The bolt flashes several times per second with about twice the power of the original strike due to the constructive interference that creates the standing LEW. The wave oscillates through the conductive bolt until the cloud's electrostatic potential is rundown. At the strike site, earth ground potential varies from zero to millions of volts.

Experiments with [43]sound waves imposed on a plates and tubes present a simple analogue of LEW behavior. Figure 4.2 shows Styrofoam beads in a plastic tube with sound injected at one end. The beads form node and antinodes that change positions in the tube as the frequency shifts. The LEW creates a similar effect on the electrons on the earth's surface. Both standing and moving waves occur in a wire or earth as a high frequency LEW progresses over the surface. Tests established the LEW exists as a moving or a standing wave. Resonant LEWs are normally standing waves with two compression points 180° apart.

---

[42] https://en.wikipedia.org/wiki/Superposition_principle
[43] https://www.youtube.com/watch?v=wvJAgrUBF4w

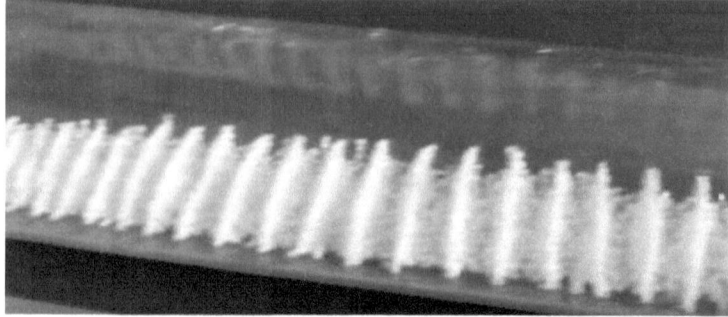

Figure 4.2 Sound Wave nodes and antinodes with Styrofoam beads

Look again at figure 4.2. Every[44] sound wave cycle has is one compression and one rarefaction wave. Styrofoam beads gather stand in a column at the nodes or zero energy points. These are termed rarefaction (*nodes*). Compression points (*antinodes*) are where there are fewer beads and they are agitated in fast motion. The medium in this case is air, and it forms the nodes and antinodes; Styrofoam beads simply reveal them. The LEW causes a similar disturbance in the electron medium.

Tesla created an experiment to reveal the soundwave property (*transmission via compression waves*) of the longitudinal electric wave. His experiment is shown in figure 4.3-A.

---

[44] https://www.youtube.com/watch?v=cBZmyG-WqNo

Thick copper bars with various nodes and conditions

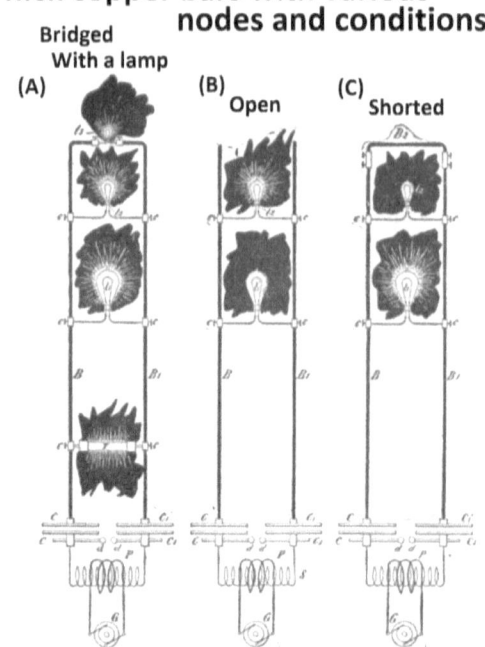

Lamps across a high and a low node point burn brightly

Figure 4.3-A Spark-gap driven longitudinal wave nodes and antinodes with three conditions

[45]All three examples use a high frequency and high-voltage spark gap arrangement for power. The spark gaps 'd', capacitors 'C', solid bars 'B', and generator 'G' make up the design. The transformer secondary supplies about 2,000 to 10,000 volts at high frequency, 10-20 kHz, depending on the generator.

[46]Example 'A': a lamp bridges (*placed across*) the top of the bars. The compression waves are moving, there are no stationary waves, and all the lamps light. The Next example, 'B' is an 'open circuit' instance, notice the center lamp lights dimly due to 'open-ended' reflection. The top lamp is brightly lit exposing the position of the antinode and node points. And finally, 'C' is a completely shorted and 'closed-ended' circuit, but one lamp lights at the center. There is only one node and antinode with the voltage difference needed to provide power to the lamp at 10kHz. Stationary waves created by a consistent frequency source are responsible for the nodes and antinodes remaining in place. Nodes and antinodes can travel or park depending on various circuit conditions, conductor length, and the frequency of the signal.

---

[45] https://www.youtube.com/watch?v=tlx7tDNXYR8
[46] Drawn by Nikola Tesla circa 1893, *labels (A), (B), and (C) and 'bridged with a lamp, open, and shorted', were added by this author*

When the High Frequency AC moves through a conductor in the manner illustrated in figures 4.1 and 4.3-A voltage potential differences appear in the conductor. Maxwell's equations (*in the Ampere's Law portion*) defines the compression wave and the current density over time with the equation [47]$J_{total} = J_{conduction} + \partial D/\partial t$. The 'D' is displacement current density and the capacitive current flow. The capacitors are the terminal spheres of the Tesla reactive power transport system (RPT). The reactive current (*VAR*) also flows in each 180° portion of a longitudinal electron wave (*LEW*).

The LEW and Maxwell's Current Density

$$J_{total} = J_{conduction} + \partial D/\partial t$$

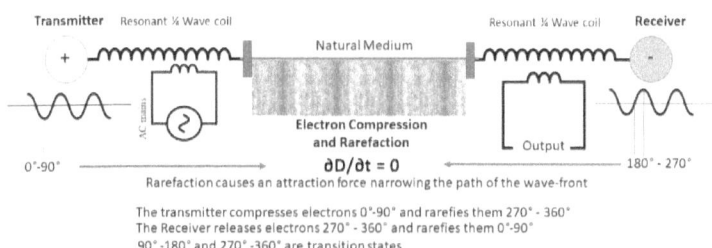

Figure 4.3-B A LEW moves to-and-fro every 180 degrees per cycle

In a Tesla oscillator $J_{conduction}$ defines current density in the antenna wire transferred by transformer action from the primary power source. In a LEW (*figure 4.3-B*), electron displacement occurs twice every cycle, at 0° – 90° and 180°-270° (*The first half of each 180° or 1π radians*) through the NM. Time (t) is the time between 0° – 180° and 180° - 360° in a resonant LEW. Transition of terminal polarity (*opposite potential on the terminal capacitor*) begins between 90°-180° and 270°-360°.

In every cycle the first 180° propagates from the transmitter to the receiver, the second 180° from the receiver to the transmitter. The 'D', displacement current, is defined by the longitudinal electron movement at completion of every ½ cycle. After a full or half cycle of 360 degrees the displaced electrons return to their original position. The sum of the AC current is zero Amperes (*D = 0*) without a load on the receiver. With a load on the receiver 'D' is asymmetric: more current can be taken from the first 180 degrees than the second. Standing wave voltage builds up between resonant cycles also contributes to asymmetry.

---

[47] https://www.researchgate.net/publication/302966559_Maxwell's_Original_Equations

The LEW is a [48]Scalar wave according to the definition: a wave that crosses the earth without (or negligible) loss of energy. [49]Tesla and Maxwell recognized these waves exist and they travel at [50]$\pi c/2$ meters/second. A scalar wave travels through the earth at 470,912,892 m/s. Tesla verified this figure at Colorado Springs. He worked out his calculation using co-tangents over the surface of the earth as shown in Appendix A. He described the movement like the movement of the moon's shadow over the sphere of the earth in an eclipse.

I proved the existence of LEWS (*Longitudinal Electron Waves*) but I cannot prove that they are Scalar waves (*Appendix B has an official description of Scalar waves*). That proof requires a very powerful, expensive Tesla transmitter. Power can be transported by scalar waves, the total energy flow between the transceivers can be defined as $\overline{S} = \overline{E} \times \overline{B}$ watts/m^2 in each direction: toward the (or antipode) receiver or away from it. I have proved through experiment that the LEW travels through the earth and water at light-speed. High frequency LEWs are attenuated over distances; low frequency waves below 100 Hz lose very little power.

A scalar wave has no time-varying B field, but if the LEW *is* a scalar wave, as I believe it is, it does indeed have a time varying E field, but not in the 'Hertz wave' sense. The LEW is a mechanical wave not bound by time changing fields as an EM wave and is therefore not confined to lightspeed limits. Mu and Epsilon naught have no velocity defining limits on a LEW as with an EM wave.

The LEW's E field is only negative, I will cover this in chapter five on electron plasma fields. Mathematically $\overline{S}$ in the previous paragraph is a 'potential' within the LEW. The potential varies in degrees of negative charge between compression and rarefaction points. The [51]Poynting theorem indicates no electrical power flow. This is exactly true. Because the LEW is a mechanical scalar wave, not an EM emission. The electric 'E' field of a LEW consists of field lines from the Ionosphere terminating on the electron medium. The 'E' field is not electrically generated by a transmitter. This completely satisfies $\overline{E} = \overline{B} = 0$ in the description of a scalar wave. An important thing to know is a LEW is a mechanical wave in an electron medium, but when it intersects any conductor in the earth it changes to normal AC power. The LEW assumes all the electrical characteristics of a normal current with electron and electrostatic power flow with normal magnetic and electric fields.

The LEW compression wave moves toward the [52]sender's antipode, whether the sender is a receiver or transmitter, every ½ cycle. Power is always taken at the

---

[48] Scalar wave See Appendix A also https://www.cia.gov/library/readingroom/document/cia-rdp96-00792r000500240001-6

[49] https://teslauniverse.com/nikola-tesla/articles/faster-light

[50] https://invention.blogflop.com/2018/03/04/regular-55/

[51] http://web.mit.edu/6.013_book/www/chapter11/11.2.html

receiver secondary winding each half cycle. In other words, during the transmitter's compression or rarefaction stroke, an electron shifts a certain distance '*D*'. In the first 180° compression cycle the electron moves '+D' and then returns to the original position '-*D*' or rest at zero volts.

During the 180° rarefaction cycle the electron moves '-D' toward the transmitter. In the second half of the rarefaction cycle the electron returns to '+D', its original position at zero volts. The net current flow from the electron movement is zero after a complete 360° or at every 180° point. The action continues with every wave. The distance the electron moves corresponds to the voltage placed on the ground plate and the bulk modulus of the electron medium. The higher the voltage the greater the electron displacement. These oscillating electrons provide the power at the receiver.

The receiver's discharging sphere and collapsing magnetic field from the ¼ wave coil reactively returns a LEW to the transmitter on rarefaction. There is an interesting characteristic caused by the positive potential on the transmitter's terminal during the rarefaction half cycle. The electron deficient ions on the *transmitter's* sphere attracts the returning LEW by rarefaction (*drawing*). The drawing force causes the LEW to form a more direct, conductive pathway from and toward the receiver (*on the compression phase*).

The LEWs are coerced into the narrower conduction path only after a standing wave condition with a receiver has been established. This aspect of LEW conductivity is a mixed blessing. It indicates several transmitters and receivers must be spaced in strategic locations to create a grid power area for aerial vehicles. It also reveals that broadcast LEWS with their circular LEW pattern are beneficial for more area coverage. The exception to this 'trough' effect is when resonance is established by antipode reflection from a powerful transmitter. The waves move in concentric circles over the earth.

Maxwell and Tesla's Reactive Power Transport

In the equation $J_{total} = J_{conduction} + \partial D/\partial t$, t is the time between 0° to 90° and 180° to 270°. The formula output power is calculated each .25 cycles and summed over time of one second of time or period. Time, t = (1/F) * .25: where F is frequency, 1 second, and .25 is ¼ of period (t) of a cycle also known as the quarter-wave. Each ½ wave of the LEW can deliver power, but not in excess of the power supply at the transmitter. If more power is used on the first or second ½ cycle than the primary power supply is providing, then the power is out of balance and the LEW transport efficiency will drop. This necessitates more input power from the transmitter to maintain the standing wave.

---

[52] A sender is a transmitting device – it can be a transmitter or a receiver responding to signals

Maxwell's original equation $J_{total} = J_{conduction} + \partial D/\partial t$ is used to calculate the VAR current in a LEW in one direction. I mentioned earlier in the chapter the $\partial D/\partial t$ portion normally equals zero because the rate of change is equal and opposite in electron movement after each full cycle (*360°*) of a LEW. In Tesla's reactive transport arrangement, each half cycle 180° represents one simple reactive circuit as shown in figure 4.3-C below.

The current delivered by the primary is the driving force and the changing current over time (t) in $\partial D/\partial t$ provides the additional transport voltage (*terminal capacitive storage*) through constructive interference of the standing wave. There is an abundance of electrons in the earth that contribute most of the VAR reactive power. Therefore, the power residing in the standing wave can be used only for transport. The sum of the reactive power in a LEW is zero, it can't be drawn off and consumed without causing an RSC (*resonant signal collapse*) condition. There is one exception: the LEW can intensify from constructive interference. Using the excess stored LEW power to charge batteries or to drive smaller loads is allowed if the LEW transport voltage is not allowed to drop below twice the voltage of the transmitter ¼ wave coil.

It is a simple circuit:

The simple circuit elements are transmitter and receiver terminals, ¼ wave coils and a common ground connection for the conduction. It is a simple earth return circuit where power is placed on a ground rod in one place and it conducts up another rod some distance away to complete the circuit, and only at one frequency. The current is AC; the transmitter and the receiver are exclusive in that they only conduct at the resonance defined by their series tank circuits. Other signals are filtered out unless they are of harmonic frequencies. The terminal capacitors work together like a battery. One terminal is positive the other negative. The polarity is switched by the AC power input on the transmitter primary. Power flows to and from terminal to terminal as normal reactive power does. It is very simple as shown below in 4.3-C.

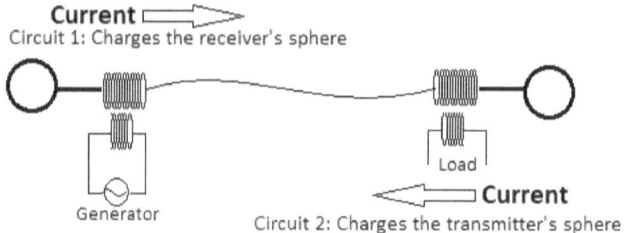

Current shuttles back and forth

Figure 4.3-C

The current moves from transmitter to the receiver representing one circuit. Returning current to the transmitter is the other circuit. The circuits are time division routes existing for ½ period of the wave. Power can, and usually is removed asymmetrically from the full cycle. The $\partial D/\partial t$ part of the equation in an RPT equals zero when the transmitter is resonating with the antipode and no power is consumed. The power from the transmitter is always greater than the returning power from the receiver. The total current used is expressed as J_{total} = $(J_{conduction} + \partial D/\partial t)_{Trans}$ - $(J_{conduction} + \partial D/\partial t)_{Recv}$. I suggest using a couple of ammeters with diodes at the receiver to determine power in and power out. In that way LEW 'in' current and LEW 'out' current can be independently monitored.

The period is 1/F where F is frequency and 1 is one second. Assume a longitudinal wave frequency of 100,000 Hz. Each wave has a period of $1/100000_{Hz}$ = .00001 ($1*10^{-5}$) sec. A quarter wave has a period of 0.0000025 ($2.5*10^{-6}$) sec. There are two quarter-wave power cycles every $1/100000^{th}$ second, one at 90° (-) and the other at 270° (+). The ¼ wave current displacement ($\partial D/\partial t$) charges the transmitter's terminal capacitor to maximum negative and positive potential one time every cycle (*100,000 cycles/sec*) to satisfy displacement $\partial D/\partial$ = 0. The total charge is a net zero (0) because reactive power is returned to the source. This balances Maxwell's equation-One (*Gauss' law*) for J_{total} on each terminal capacitor transmitter and receiver at the 180°- and 360°-degree points of every cycle.

The terminal capacitance defines transport power

A simple 20-inch spherical terminal with no dielectric covering is considered. A sphere with this diameter has a capacitance of 2.8 nFd ($2.826*10^{-9}$) where C = $4\pi\epsilon_0R$. One Farad is defined as a charge (Q) of 1 coulomb (*1 ampere*) at 1 Volt (V) in one second. Our terminal capacitor can contain a charge Q of .0000000028 coulomb per volt per second. The terminal will charge and discharge 100,000 times per second. The polarity reverses (*every 1/2 wave from – to +*) 200,000 times per second. The negative charge of the terminal capacitor at 100,000 times per second (*0°-90°*) is 0.0000000028 Coulombs/s * 100000Hz = - **.0028 amps per second** at **1 V** negative; the ion magnitude is the **same** for the positive charge. The total current each cycle of the terminal at 1 volt is 2 * 0.0028 = 0.0056 amperes for one charge and one discharge (*peak-to-peak - one positive and one negative per ½ cycle*). The terminal will be charged the total sum of 560 coulombs (*Amperes*) per second at 100*10³ Hz at **1 V**, assuming the RC time constant is such to allow the charge and discharge rate.

The quarter-wave coil inductor

A small transmitter with a primary current of 200 Amperes at 100,000 Hz at 100V peak-to-peak is a small (*20,000 W*) Tesla RPT oscillator. Each ½ wave has a current capacity of 100 Amps per second at + or - 50 Volts. Assume the oscillator's secondary winding has a turns ratio (*TR*) of a 260:1 voltage gain, a common ratio for a Tesla transformer. The voltage and current transformed equals [53]260_{TR} * 50V (1/2 wave) = -13,000 V at .38 amperes per second ($100_A/260_{Tr}$) every half-wave; the Tesla RPT pair transfers power every ½ cycle. A full cycle is 26,000 Volts peak-to-peak and .769 Amperes per second taken from the power source of the transmitter's primary by transformer action. At .769 amperes per volt per second calculation confirms .769A/s * 26,000V = 20,000 Watts full-wave amperes (not counting a 5 percent transformer loss) from the primary power source at 100,000 Hz.

The voltage figure of 26,000 Volts is doubled or tripled by the resonant power of the standing wave. The voltage is reactive and the LEW current displacement ($\partial D/\partial \approx 0$) is zero after each cycle. The standing wave is a 'for free' 180° power transport system. Current can be removed at the receiver every ½ cycle through transformer action as shown in figure 4.3-B. The high frequency currents can be rectified and inverted to the needed voltage and frequency. Power can also be removed at the receiver through capacitive coupling on the terminal capacitor as shown below in figure 4.3-D. Grounding one side of the load is not necessary but it is more efficient.

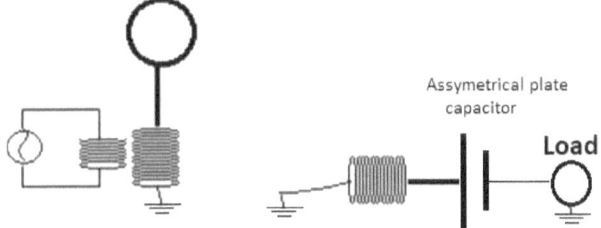

Method of taking power capacitively from a receiver

4.3-D A capacitive coupled method of taking power from the receiver

---

[53] TR is turns ratio

An abundance of power is stored in the standing compression wave. The power is usable, but it can quickly deplete in a few seconds from over-drawing at the receiver. This leads to an RSC (*Resonant Signal Collapse or standing wave failure*). The power in the resonant wave is provided by the primary power source at the transmitting oscillator. The standing wave fluctuates the electrostatic lines of force from the ionosphere terminating on earth electrons. The fluctuating electrostatic field's power can be collected by a proper receiver, providing an abundance of power for aerial machines such as cars, phones, and drones, see chapter five.

In the equation $J_{total} = J_{conduction} + \partial D/\partial t$, the $J_{conduction}$ is the primary power supplied at the transmitter and is the actual usable power transported in the LEW. The standing resonant wave is an efficient 'free', power transport mechanism providing the benefit of wireless power transmission of J_{total}, but at the risk of being to repetitive, it is reactive power meaning it leaves from and returns to the transmitter each cycle.

Making sense of the 'power gain' of the Tesla RPT:

In the oscillator described above, actual current capability of the primary power source is limited to 200 amperes per second. This amounts to the primary power supplied at the transmitter. The only 'free lunch' gotten from the Tesla wireless power system is the 'free' transport and the Electron Plasma Waves caused by LEW agitation of surface electrons.

The standing wave LEW multiplies in strength with each oscillation. The high voltage can cause a lightning type of surge to erupt from a transceiver's coils or terminal. The high voltage from resonance carries extra power (I^2R) because earth electrons are conducted up to the terminal and added to those of the copper. Terminal voltage should be monitored and kept within acceptable levels by reducing power into the oscillator primary or using the excess LEW power to trickle charge batteries. This conserves some of the energy that would otherwise be lost to sparks and capacitive power drain. Another dangerous aspect of the resonant LEW is power feedback at the transmitter. The high voltage is reversely induced into the primary winding. This destroys power suppls, burns-out transformers, and kicks out circuit breakers leading to an RSC event. Tesla experienced this in Colorado. He had to pay the power company for repairs on their generator.

Transmission of power:

Standing-wave antinode (*compression*) and node (*rarefaction*) points vary in number according to the frequency of the currents. The coulombs (*Amperes*) of electrons moving in the medium also vary as the voltage and frequency driving them. The shorted bar example 'C' (*Fig. 4.3-A*) has less than .005 ohms resistance, yet the lights will glow when they connected at the node and antinode point on the copper bar. These three examples are common with radio frequency currents. LEWs also have these nodes and antinodes with a potential difference between them just as seen in these examples.

Seeing wave movement in a bar of copper is helpful toward understanding LEW movement through the earth. Many factors complicate the earth conductor and LEW transport. There are LEW Attenuation Spots (*LAS*) in the NM where electrons are not as compact. This presents a unique set of challenges for LEW transport. The dead spots can act as a terminal end and reflect the LEWs back to the source.

System engineering cannot account for all the unknowns. LAS locations and their effects are resolved individually. There are no 'LAS tables' or charts to help in the design phase today because there is no world-wide RPT system. When and if the RPT system becomes common, a LAS table can predict LEW density and repeaters can be positioned in the deficient locations.

The LEW does not move through the earth like electricity moves in a closed circuit. The LEW is an induced-potential voltage, mechanical-style compression wave traveling through the earth like a TDR (*Time Domain Reflectometry*) pulse travels through a wire. A TDR tester induces high voltage pulses at an end-point in a wire. The pulse wave moves to the other end of the wire and reflects to the transmitter like light off a mirror. The wire length is defined by the return time of the pulse with the additional consideration of temperature, wire gauge, insulation, and metal type. The earth is an excellent conductor (*.0055 Ohms*) with a very low resistance. A high voltage signal induced into the earth ground plate by a Tesla RPT transceiver will travel directly toward the geographically opposite point on the earth (*antipode*). The signal echoes back to the transmitter just as a TDR signal reflects in a wire. The cadence of the LEW pulses traveling and rebounding in the earth constructively interfere and create a [54]standing [55]longitudinal wave. This standing wave rhythm is LEW resonance.

Looking deeper at the LEW

---

[54] https://en.wikipedia.org/wiki/Standing_wave Definition and animation of a transverse wave
[55] https://www.youtube.com/watch?v=0f5iYCNCnow
  http://hyperphysics.phy-astr.gsu.edu/hbase/Waves/standw.html longitudinal examples

LEW wave traveling in the NM

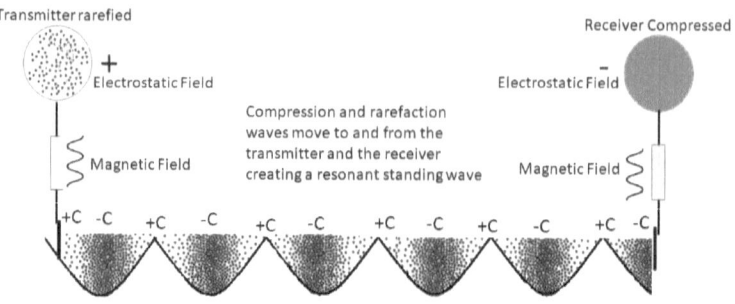

**Figure 4.4-A The Standing Longitudinal Electron
Wave in the Natural Medium**
Rarefaction points charges are +C and compression are -C.
The terminals are charged oppositely.

In a Tesla Reactive Power Transport scheme, the receiver mimics the antipode of the resonant transmitter. The Tesla resonant pair forms an intensified standing wave pulsing negatively through the earth two times for every three-hundred-and-sixty-degree (*2π*) cycle of the transmitter's quarter-wave winding. Proper [56]impedance matching between the oscillators boosts power transfer efficiency. Interestingly, the point is to mimic a low (*terrible*) [57]ERL (*echo return loss*) and it done by perfectly matching impedance between the oscillators. The poor ERL is caused by the electric discharge of reactive components back into the NM in a resonating 'feed-back' arrangement.

The standing wavelength pattern varies according to the distance between the transceivers (*time for the signal to echo*) and the resonant frequency. The Tesla resonant transceivers function by signal feedback where [58]ERL is as low a ratio as possible, near zero dB or almost no loss in power from the original signal. The oscillators squeal at resonance. At low frequencies the feedback is audible like a sixty-cycle hum in a transformer.

LEWs are simply 'sound' waves on the 'sea' of earth surface electrons. In 4.4-A, the charge (*C*) of LEW points are labeled [59](-C) for compression and (+C) at rarefaction points. The sum of the voltages in a full cycle of a resonant LEW is negative. The highest voltage magnitude is in the compression points of the wave. The rarefaction points are zero volts and they are the positive potential of the LEW.

---

[56] https://en.wikipedia.org/wiki/Impedance_matching
[57] https://en.wikipedia.org/wiki/Return_loss
[58] **Echo Return Loss** (ERL): is the ratio between the original signal and the **echo** level expressed in decibels (dB). Simply put, ERL is a measure of the **loss** of the signal that comes back as **echo**.
[59] +C and -C are negative and positive Coulomb charges in the LEW

For every three-hundred-and-sixty-degree cycle (*or full wave*) of the primary coil power at the transmitter, the LEW changes direction twice.

The reason for two negative strokes

The first (180°) half of the power input sine wave drives the transmitter's ground plate negative (*figure 4.4-C*) repelling electrons. The negative potential at the ground plate forms a compression wave that moves outward in concentric circles toward the antipode. When the LEW compression wave reaches a resonant receiver's ground plate, it conducts up the quarter wave coil and charges the receiver's sphere. If the 'DNA' matches (*the resonant timing is perfect*) the receiver will resonate with the transmitter. Otherwise the series resonance doesn't allow conduction to the sphere and the power is returned to the NM usually unused (*in most cases*). This is true of any non-resonant conductor charged by the LEW.

During the second half of the power stroke (*180°-360° degrees*), the resonant receiver unloads its electrons back into its earth plate, echoing and mimicking an antipode. The LEW from the receiver passes through the NM and enters the transmitter's ground plate. It continues up the quarter-wave coil and recharges the transmitter's sphere in synchronization with the driving power signal of the primary. This action causes two negative pulses into the NM every full AC cycle, each one-hundred-eighty degrees apart. The resonant LEW is always negative because of the oscillating (*negative charges*) electron medium. As I mentioned earlier, the rarefaction cycle at either transceiver causes a conductive trough between the two oscillators. The LEW still spreads, but a LEW power concentration path (*usually the shortest distance*) is formed between the resonant oscillators.

Figure 4.4-B below illustrates electron displacement of the LEW compression wave caused by the voltage across the primary winding powering the oscillators. The two dots on the vertical sine wave in the figure represent one electron in the medium. The voltage amplitude induced into the quarter-wave coil by the primary winding expresses the power of the compression and rarefaction cycles (*the LEW*). The resonant frequency of the quarter-wave coil and geographical separation defines the number of the LEW's standing wave nodes and antinodes (*the number of times per second the compression wave is oscillated through the medium*).

Signal voltage amplitude determines the
reciprocating displacement of each electron
in the compression wave

Total compression displacement =
Peak-to-peak voltage of the input
power

Figure 4.4-B Compression wave movement through the NM

[60]Compression waves move to-and-fro between the transmitter and receiver see figure 4.4-C. When the compression waves meet, they combine [61]constructively doubling   in   energy (*electrical power*). A standing wave pattern forms because of constructive interference. The forward moving waves and the returning waves are the same frequency in resonant LEWs.

The standing [62]LEW pattern in figure 4.4-C is defined by $L_{an} = d/(\lambda/2)$ where L_{an} is LEW standing wave antinode pattern frequency, d is distance between transceivers, λ is the resonant frequency (*wavelength*) which is divided by 2 defining the ½ wavelength of the LEW. There is always a half-wave stored within the transceivers and is not considered.   The equation does not calculate the fundamental harmonic, but the number of half wavelengths between the transceivers. For example, assume a 100,000Hz resonant signal with a distance between transceivers of 50 miles.   First the wavelength λ = $186000_{mi/sec}/100000_{cycles/sec}$ = $1.86_{mi/cycle}$.   Apply the equation $L_{AN} = d/(\lambda/2)$ = 50/(1.86/2) = 53.7 antinodes at 180° (*1π radians or 1/2 cycle*). The wave pattern in figure 4.4-A shows how the standing LEW would appear if it could be seen. The individual antinode length is $50_{mi}/53.7$ antinode = $.93_{mi/antinode}$.   The perfect even number of 53 or 54 antinodes can be set by varying the distance between the transceivers, varying terminal capacitance, or adjusting the length of the

Standing Wave LEW

The medium is the electrons on the earth's surface
therefore all compression points are negative

Transmitter   Receiver   Antipode

Reflection

Direction of travel

Dark bands are compression   Light bands are rarefaction
points which are negative   points which are positive

90°   90°   90°   90°

The primary's driving sine wave

transceiver's quarter-wave coil and/or ground wire.

[60] https://www.youtube.com/watch?v=HIN0d38Q_WY (standing compression waves)
[61] https://www.britannica.com/science/constructive-interference
[62] https://youtu.be/0yEtLeqUlEQ Video showing actual LEW transmission using a low-power signal generator

Figure 4.4-C The standing wave composed

of traveling LEW compression waves

The result of 53.7 antinodes between the transmitter and receiver indicates that .7 of 180° cycles remains; Theta (θ) = 126° (derived *by 180° * 0.7*). To achieve maximum efficiency of an even 54 antinodes, the remaining fraction of the LEW .3 mi (*53.7 + .3 = 54 miles*) of the compression waves must be split between the transmitter and receiver wire length. The length to be added or removed from each is (λ/2 * .3)/2 = ((1.86/2) * .3)/2 = .279/2 = .14 mi or about 737 feet each. The object is to achieve a maximum peak resonant standing LEW by varying the ¼ wavelength coils.

The simpler method to change frequency is done by adjusting the terminal capacitor and changing the input voltage to match the coulomb capacity required at the receiver/s. This method has drawbacks because of the complications arising from re-engineering the power delivery at any given frequency. This problem can be a challenge when steps are taken to prevent power theft. Encryption by frequency hopping, OFDM signal generation, or time division multiplexing (*TDM*) is a good strategy but it is difficult to engineer. Harmonics of the fundamental frequency are easily determined by would-be thieves, so frequency hopping, within harmonics of the fundamental frequency is not secure. TDM, and [63]stereophonic transmit schemes can be used in the transceivers. Transceiver frequency control may use sideband communication to establish the encryption on the fly.

There also can be multiple transmitters and receivers in each arrangement. Mixing signals with only three such transceivers in each location can generate thousands of combinations of frequencies. The signals mix at the ground plate of the transceivers into a single transformed frequency pattern. The good news is that it isn't necessary to have a perfect standing wave. The transceivers will transfer power efficiently without them. The importance of a resonant standing wave is for modulating the plasma field between the Ionosphere and the earth's electrons for powering aerial vehicles. If aerial vehicle power is not part of the plan then geographical separation planning is not necessary but resonance between the oscillators *is* still required.

More about the LEW

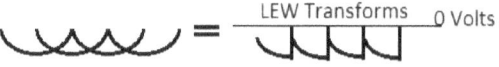

Complex LEW form

---

[63] https://en.wikipedia.org/wiki/Stereophonic_sound

Normally sum of the voltage in rarefaction points is zero volts unless the LEW is [64]Complex. The LEW is a sound-wave in an electron medium. Launching a LEW is nearly identical to creating sound waves with a drum; LEW compression waves are simply disturbances in the electron medium. The standing LEW mimics the standing sound wave which is responsible for the loud, squealing feedback noise heard in auditoriums during sound testing. Energy grows with each cycle of the wave over time; this is resonance. In figure 4.4-D the sound power compression and rarefaction points are 90° apart, not 180° like the LEW. This is the big difference between sound and LEW waves.

The resonant LEW has three nodes (*0-Volt points*) and two antinodes (*negative Volt points*) per cycle (*fig. 4.4-A*). The medium is electrons (*negative charges*) and current does not flow as in a normal closed loop 'Kirchhoff' type circuit. The Tesla reactive oscillators charge and discharge capacitors on either end of a 'single wire' circuit as mentioned earlier in the chapter. The one-wire circuit is from terminal to terminal through the earth (*the one wire*) driven by a primary AC source. Power shuttles between the capacitive spheres.

In wireless power terms, Nodes (*rarefaction*) and Antinodes (*compression*) are points where zero and maximum voltage exists in a resonant LEW. If you glance back at figure 4.1 and notice, the lamp is across a node and antinode where the highest voltage potential exists. The lamp glows because the wire terminals are at these strategic points where power pickup points are 180° (*not 90°*) apart because this is a closed-circuit with full 360° waves. The generator's frequency defines the number and location of the antinodes and nodes. More of them appear as the frequency increases.

The two types of LEWs

LEWs come in two types, full-wave and half-wave. The **full-wave** LEW (*fig. 4.4-D*) is a non-resonant, non-standing compression wave with positive (*not zero*) antinodes and negative antinodes located 180° apart. Rarefaction points have an electron deficit and are positive (*not zero*), the antinode compression points are negative since the medium is [65]electrons. The medium has a negative charge, and a super-high bulk modulus. The power-driving voltage at the ground plate can be pulses (*negative or positive*), square waves, or sine waves. The electron medium around the ground plate compresses or rarefies making dense, powerful waves causing electrostatic charges and a magnetic field at the earth's surface. And like I mentioned earlier LEWs are like TDR current traveling in a wire. The transmitter can be as simple as a signal generator or as complex as a Tesla coil.

The LEW seems to defy the definition of standard electric current. The Longitudinal Electron Wave travels through a medium of electrons moved by

---

[64] A three phase Longitudinal electron wave created by an array of transmitters not within the same ground plane.
[65] interstitial electrons attracted to the earth's surface by the Ionosphere

powerful currents at the ground plate of a Reactive Power Transceiver. By [66]definiti on LEW movement is an electric current. Energy transports through wave pulses, electron movement is only oscillatory and equal to zero volts at the start and end of every cycle. All the standard properties of electrical AC current do apply to the LEW. Each half wave is considered a pulsing DC circuit with a positive and negative terminal.

A Full Wave Non-Resonant LEW

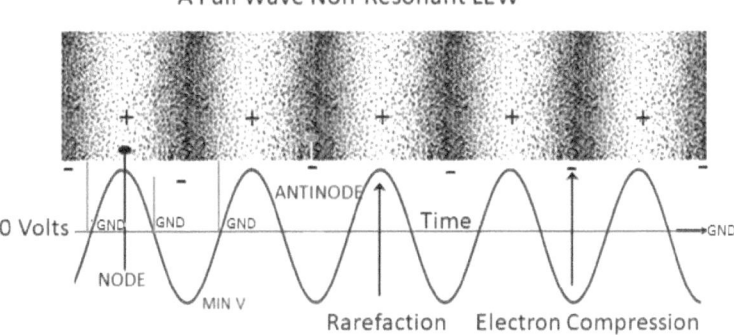

Figure 4.4-D A non-resonant, non-standing full cycle LEW. Nodes are rarified and antinodes are compressed points.

The voltage sum of a non-resonant, **full-wave** (*360°*) LEW equals zero from a peak-to-peak. Full power can be taken from peak to peak, at half voltage points between an antinode and node, or at 90° points in the LEW. However, a resonant LEW has 0 volts and negative voltage points, with no positive points in the wave.

This is initially confusing. The resonant LEW has at least *twice* the voltage in each 180° part of a cycle or four times the power in a full 360° cycle. The voltage is always negative and the sum of the voltage from peak-to-peak is four times the absolute value of the input voltage times -1. The voltage amplification is a result of resonant, constructive interference.

Constructive power in the LEW

Referring again to figure 4.4-A, the wave compressions are negative and rarefactions are zero volts. Assume a supply of 20,000 Volts peak-to-peak were applied on the transmitter quarter wave secondary. This causes a negative 10,000 Volt compression wave and a positive 10,000 Volt rarefaction wave like the full-wave LEW. Rarefaction cycles (*from a positively charged transmitter terminal*) draw electrons up the quarter-wave coil of the transmitter from the earth along with the returning negative compression wave from the antipode or resonant receiver. The cadence is at the resonant frequency of the oscillators.

---

[66] https://en.wikipedia.org/wiki/Electric_current

A resonant LEW is a [67]standing wave with power-doubling constructive interference effects which occurs when the LEW of the transmitter collides with a returning LEW from the receiver.

Examining a full cycle (360°) between established transceivers

First, the transmitter's power supply injects a negative signal of -10,000-volts into the quarter-wave coil and its ground plate. The resulting compression wave moves away from the ground plate toward the antipode. **Second**, the wave encounters an a priori LEW from the resonant receiver. There is a series of such waves already traveling in the medium and their number depends on the frequency and distance to the antipode or the receiver. **Third**, when resonant waves collide, they add constructively because they are the same frequency and polarity. The interference causes the amplitude to double (*-20,000, -40,000, etc.*) volts every ½ cycle until the energy is consumed or released in some way.

Fourth, the next 180 degrees (*final cycle 180° - 360°*) of the transmitter causes a rarefaction in the terminal and the receiver's compression wave is released at the same instant because it is in resonance. **Fifth**, the negative compression wave of the receiver (*released into the ground plate by the static discharge of the receiver's sphere and collapsing magnetic field*) moves toward its antipode; the transmitter ground plate intercepts it. **Sixth**, during the compression wave's travel, it collides with an incoming wave launched earlier from the transmitter causing the LEW to double in voltage.

Review

The wave front collisions cause a pattern of standing waves to form with -20,000 volts or more in each 180° part of the cycle. The wave train moves back and forth two times per cycle. A LEW of -40,000 volts will double again and again and the amplitude can get out of control and feed back into the power supply and the receiver's inverter destroying it.

Full wave LEWs exist only when the signal **is not** resonated or terminated and dies out over a distance. These LEWs types are low-voltage with low frequency, or high-voltage with high frequency. For example, a LEW of one MHz at ten million volts will fizzle out in a few miles. The earth's inductance and other attenuating ground interference defines the range. If a LEW cannot resonate because of attenuation (*not enough power to reach the antipode),* it classified as a one-way, receive-only, or a broadcast LEW.

Diurnal effects on LEWs

---

[67] https://physics.info/waves-standing/ Video - https://www.youtube.com/watch?v=NpEevfOU4Z8

Cosmic rays, sunlight, and heat can cause increased electron energy excitement on earth-surface electrons. This in turn slightly changes the electrons bulk modulus causing longer wavelengths and standing wave pattern changes. Depleted electron density and atmospheric chemistry will be covered in chapter five. Other causes of wave length variance are intermodulation distortion, transmitter primary harmonics, or harmonic interference in the resonating receiver. [68]Electrons have mass and are incompressible. But the 'E' field (electrostatic field) surrounding them can be compressed. Highly excited electrons are less conductive, 'spongy', and offer less power transport capability.

In chapter three I touched on the electrical properties of the earth. The sea of electrons at the surface are interstitial, attracted to the surface by positive electrostatic forces. [69]Positive lines of force from the Ionosphere connect with each surface electron. The LEW causes undulations in the 'electron sea' and the field intensifies at LEW compression points. I call the lines of force an electron plasma field (EP field)

When the LEW moves over the earth, the compressed electrons nudge the electrons next to them in a push/pull fashion. This continues in a chain reaction toward the antipode. When the LEW moves through the surface a high frequency magnetic field (twice the LEW frequency) and an intense downward pointing vertical (a sink) 'E' field follows them. LEW movement through the NM modulates the 'E' fields. The electric field forms a cold plasma mixture of loose ions (+) and negative (- or sink) electrostatic fields.

The electron Plasma field (EP)

The field density is expressed as divergence (D) in [70]Maxwell's equations:

$$\int_S D * dS = Qenc$$

: where 'D' is electric field Divergence, Q_{enc} is charge encircling the sphere, and 'S' is the surface. Implying the field density is integrated over the whole earth surface, including the voltage waves in the LEW compression wave. A detailed description of EP fields will be covered in chapter five, and how they can be used to power aerial vehicles.

The magnitude of the voltage in the LEW is the same as it spreads over the sphere. The LEW charge density 'D' decreases as it moves because the same charge is spread over a larger area. It is weakest at the center-point (transmitter's equator) between the transmitter and its antipode. LEW density increases again as the wave moves away from the center-point toward the antipode or transmitter. The sum of the voltage in the LEW is constant but the wave has less 'D' (density) due to the larger circumference at the equator.

---

[68] 9.10938356 × 10 kilograms
[69] https://www.physics-and-radio-electronics.com/electromagnetics/electrostatics/electric-lines-of-force.html
[70] http://www.maxwells-equations.com/gauss/law.php

Transport

Reactive Power Transport consumes no power except for small frictional losses. The energy used to produce the LEW at the transmitter is returned from the resonant receiver. The electrostatic terminal capacitor and the magnetic field of the coils are the reactive parts. Power is taken from the receiver by transformer action in the secondary winding of the quarter wave, resonant coil or from the terminal via capacitive coupling. Excessive power drain (*power removed exceeds supply at the transmitter primary*) will cause LEW power depletion and resonant link failure (*RSC*) covered in chapter seven.

Capacitance and inductance are vital parts of power transmission systems of all types, even DC power supplies. Modern power grids would not work without reactive elements to store and return energy from the grid to the generator. Tesla incorporated these facilities to take advantage of the 'free' transport they provide for both wired and wireless networks.

Exploiting the nature of capacitance, inductance, and the NM is an efficient, energy conserving, means for power transport. Sometimes functionality is more important than costs, an example of this is the broadcast LEW. It transports power and broadband service without resonant, energy conservation. These style LEWs broadcast only for a few miles and can be used to transport power from cell towers to charge cell phones or to transmit broadband communication. Broadcast LEWs can be received by switched MOSFET or reactively designed receivers. Two-way secure broadcast (*like radio*) and device charging are feasible. Broadband Internet service is supported by resonant or non-resonant aerial receivers. Non-resonating MOSFET broadcast transceiver models can be used for short ranges depending on frequency.

Defining wave length

Tesla Reactive Power Transfer (*RPT*) assemblies resonate at a frequency defined by combination of several parts. The sphere's capacitance, the quarter wave coil length, its resistance, and its inductance. The [71]time required for electrical pulses to pass from the ground (*quarter-wave transit time*) to the terminal sets the transceiver's base frequency. For example, assuming electricity travels at light-speed, a quarter wave coil 0.3 miles long has a basic wavelength of (*0.3 * 4)* 1.2 miles. With a 1.2-mile wavelength, the frequency and period of the quarter wave coil is 155 kHz (*186000$_{mi/sec}$/1.2 $_{mi}$ = 155,000 Hz*). The period is about 6.5 micro seconds (*1s/155000 ≈ 6.5*10 $^{-6}$ s*) per wave. Quarter-wave transit time is 1.6 micro seconds (*6.5μs/4*). The half-wave signal (*180°*) has a period of 3.2 μs which is one resonant LEW power cycle.

---

[71] The time period of the quarter wave coil is determined by other factors also such as wire type, insulation, and other.

A machine with a resonant frequency of roughly 155 kHz has a standing LEW pattern of 310 kHz (2* *155000 = 310,000*) or two LEW compressions per cycle. For every ½ cycle (*.6 miles*) a compression and rarefaction point exist. The nodes and antinodes are 792 feet apart (*5280'/mi * .3 mi = 1584': 1584'/2 = 792'*).

The terminal capacitor is the driving force of the Tesla oscillator. The single plate terminal is a silo to contain electrons and ions under high-voltage pressure. The power in the standing-wave LEW is directly linked to the product of terminal voltage and coulombs of electrons stored and released. The power delivered by a LEW increases with the frequency. But higher frequencies impose a compromise between the inductive reactance (X_L) and transport range at a given voltage.

Adjustments to the quarter-wave wire length, terminal capacitance and/or quarter-wave coil inductance are used to vary the transceiver frequency. The endless supply of earth electrons for the Tesla oscillator permits scaling to any size. Ongoing constructive interference magnifies power at compression points in a standing LEW. But quarter-wave coil resistance, arcing, and electric field leaks drain the power and help maintain LEW integrity up to a point.

The RC time constant for the oscillator:

The resistance of the quarter wave coil, and the terminal capacitance define the minimum charge time for the terminal capacitor. The terminal's minimum charge time is the RC time constant (t = R*C) between the ground plate and the terminal: where 't' is time in seconds, 'R' is linear resistance, and 'C' is capacitance in Farads. The RC time constant must be much less than the period of the resonant LEW. The Q (*charge*) that can be placed on the terminal capacitor per cycle *decreases* with a higher time constant.

A terminal capacitor of 60 pFd (*.000000000060 Fd*) and a quarter-wave coil with 500 meters of #12 wire (*.00521 Ohms/m*) of 2.6 Ohms resistance has a time constant: t = RC = 2.6 * .000000000060 Fd = $15.63*10^{-9}$ seconds. The frequency of the period translates to: 1 Sec/.00000001563 Sec = 64 MHz   This is ample time for full terminal charge of any Tesla oscillator terminal operating below one MHz.

Increasing mutual inductance between the primary and secondary (*close-coupling*) allows the oscillator to be driven at multiple frequencies with high secondary voltage. Closely-coupled machines can operate at higher or lower harmonics of the resonant frequency. Transport efficiency drops with weaker LEWs at harmonic frequencies. Tuning higher or lower causes a loss of resonance and initiates an RSC (*Resonant Signal Collapse*) between harmonics. A variable frequency power supply at the transmitter provides any frequency wanted. High frequency in closely coupled machines increases inductive reactance (X_L) in the NM and in the transformer's windings and lower frequencies may cause arcing in the windings of the quarter wave coil.

One advantage to using closely coupled oscillators is for resonant testing between the transmitter and various receivers or to ascertain power at various harmonics. A closely coupled transmitter can assist in trouble-shooting resonance problems. The transmitter can be forced to send higher or lower frequencies to 'find' the receiver. I have used this method often with new receivers. It is especially useful when winding counts are wrong or for fine tuning the receiver. The receiver is tuned simply by adding wire and/or turns to the quarter-wave coil or changing the terminal capacitance. There are simpler ways to test through the NM and the instructions are in chapter seven. The method requires only a 5 V waveform generator and some other passive equipment to launch LEWs.

Determining the base frequency of a quarter-wave coil.

Every quarter-wave coil has a base frequency. The variations are caused by parasitic capacitance or coil winding errors. The planned and the actual base frequencies must be as close as possible. Dissimilarities are detected by doing a base frequency test. This test examines the resonant gain across the transmitter's quarter wave coil at various frequencies. An arbitrary waveform generator is used as the power source on the primary.

The purpose of the base frequency test is to learn the exact resonant frequency of the quarter-wave coil without a terminal or ground. Determine the transceiver's resonant frequency with a frequency sweep using the waveform generator across the primary winding and observing the highest voltage reading across the ¼ wave coil. The oscilloscope (*preferred*) or volt meter can be connected in two ways. Directly across the terminal and ground wires or by winding a simple tertiary coil of thirty to fifty turns around the quarter-wave coil and measuring the voltage across it. *If tests are done between the terminal and ground wires, high-voltage oscilloscope or meter leads are needed.* The secondary output from a five-volt waveform generator placed on the primary can step-up to 20kV on the secondary. I ruined a new oscilloscope when I did this test the first time. Do not connect the quarter-wave coil to any ground or terminal capacitor for this test. If the frequency of the quarter-wave coil has the highest gain at the wrong frequency, examine the coil for winding problems. Minor frequency flaws are corrected by adding or subtracting windings to adjust the coil's length.

High power demand

The Tesla Reactive Power Transport (*RPT*) method uses standing wave power of the LEW as an energy silo. Consumable energy can be stored for a few seconds. Energy stored in the LEW can be one hundred times the input power at the transmitter. Dual power schemes (*figure 4.5 below*) can be used in situations of high-power demand. Pairs of RPT transmitters working together can provide more usable power capacity for industrial use and/or powering cell phones, automobiles, underwater and aerial drones. Any receiver without a ground connection is an considered an aerial receiver (*more in chapter five about aerial receivers*). Dual power oscillators provide a dense LEW wireless power grid.

Dual Power System for powering vehicles, cell phones, and drones

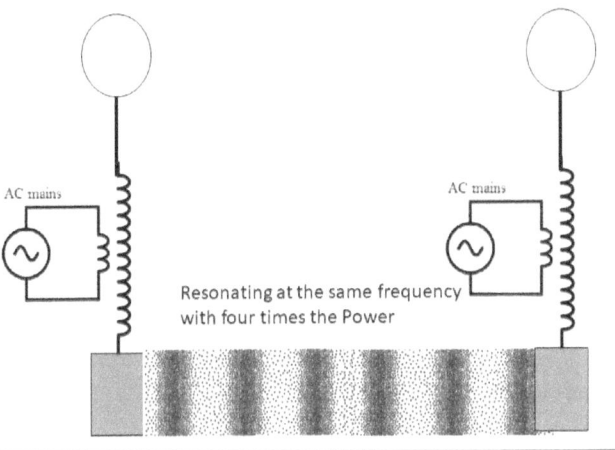

Figure 4.5 Dual power system

The velocity of the LEW:

Tesla calculated the [72]speed of the LEW through the NM. He developed the formula (πc/2), where c is the speed of light.   The equation defines LEW (*mechanical*) compression wave over a sphere and not through wire.   There are physical differences in the conductors and different mathematics apply to spheres with surface conduction.   The LEW has the characteristics of a scalar wave as mentioned earlier. At the origin the speed is infinite but gradually decreases until it reaches the equator (*of the origin*). At the equator a LEW moves at light speed and increases again becoming infinite at the antipode. The sequence repeats in reverse with the echo wave. Tesla calculated the LEW speed is greater than the speed of light from the poles to the equator region, and slows to light-speed at the equator/halfway point. This is possible for scalar waves. We need a Tesla Colorado Springs or a Wardenclyff size machine for physical proof. The scalar portion does not apply to the ¼ wave coil and the spherical capacitor, these are standard electrical components with reactive characteristics.

---

[72] https://teslauniverse.com/nikola-tesla/articles/faster-light - https://teslauniverse.com/nikola-tesla/patents/us-patent-787412-art-transmitting-electrical-energy-through-natural-mediums

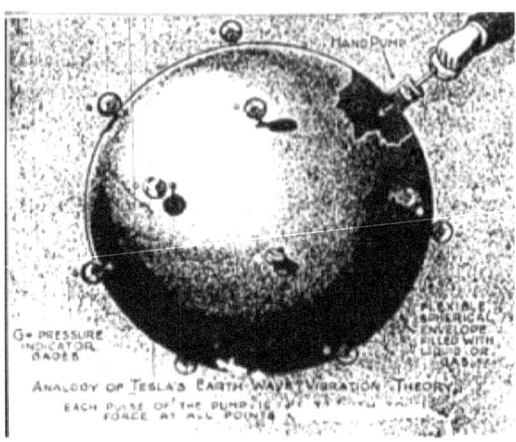

Figure 4.6 Tesla's example of a LEW

A Closer Look at reactive power transfer

Figure 4.6 is an analogue example that Tesla drew showing a fluid moving in a spherical cylinder. In this example, the hand pump represents the ground-plate of the Tesla RPT transmitter's quarter wave coil to the NM (*ground*). The fluid inside the sphere represents the earth's electrons and the expanding or contracting bladders represent the resonant, reactive receivers. The LEW style compression and rarefaction waves move in the sphere conveying power from and to the pump. There is no analogue to describe electron plasma (*EP*) and magnetic fields that arise from electron (*fluid*) movement.

Like the hand pump example, the LEW compression wave starts from the ground plate when electrons are injected into the NM by a (-) pressure stroke from the primary winding. The expanding bladders (*receivers*) are representative of the terminal capacitors. The Elastic bladder symbolizes the sphere holding electrons under voltage pressure. You can visualize how the bladders return energy to the hand pump as a pressure wave behind the piston withdrawal. The sole movement of energy in the analogue is the piston's downstroke. The elastic bladders (*reactively*) return that energy back to the cylinder. The terminal capacitor does the same but with electrons rather than fluid. An RPT receiver creates compression waves during the discharge of its terminal capacitor. The electron rarefaction and compression power vacillate between the receiver's and transmitter's terminals in the LEW (*longitudinal electron wave*). Electrical power on transmitter's primary coil is the prime mover in the analogy of the Tesla 'hand pump'.

Review

A LEW standing wave is a result of an antipode echo or a resonant receiver discharge. New outgoing waves collide with returning ones of the same frequency.

Constructive interference occurs on each colliding wave-front. The wave intensifies (*doubles*) in stored power. When the waves collide and form a pattern called a [73]sta nding stationary wave. Within one or two seconds the LEW can strengthen a hundred-fold according to frequency, electrical friction, or power drain factors. Power input at the primary winding in the transmitter maintains the standing pattern.

The step-by-step process of LEW generation:

When the electric power is applied to the primary coil of the RPT transmitter (*figure 4.5-B*) earth electrons immediately respond. Within a second a standing-wave LEW pattern forms. Depending on frequency and distance between transceivers the pattern can have many LEW power points (*antinodes*). The fundamental harmonic of the resonant transceivers delivers more power. Higher frequencies offer less range but denser power transfer and higher bandwidth broadband.

The order of events: The following scenario is a sequence of events required to establish a standing LEW.

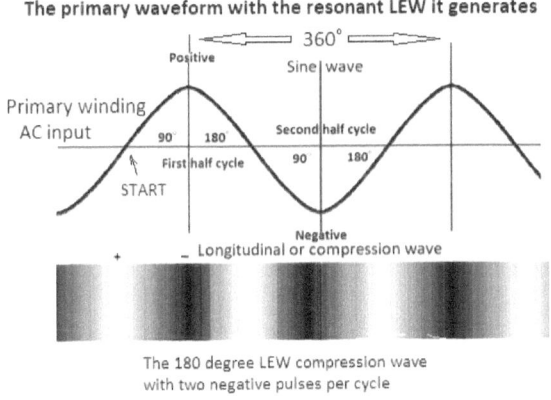

Figure 4.7-A The resonant standing LEW and the driving power Sine wave

System flow for the second harmonic and its 180° standing wave:

Step   **The power is switched on at the transmitter to the primary winding, see the START arrow '4.7-A'.** *The first stroke is assumed to begin on the (positive) 90° of a cycle as shown in figure 4.7-A.*

1.   The Sphere is like a single pole battery that switches polarity and follows the voltage of the ¼ wave primary coil. The opposite charges cause current to

---

[73] https://www.physicsclassroom.com/class/waves/Lesson-4/Formation-of-Standing-Waves

conduct through the earth, up the ¼ wave coil, and to the spheres.

2. The primary power begins the first quarter of cycle (0°-90°) placing a positive potential on the sphere Figure 4.7-B, and placing a negative potential on the ground plate.

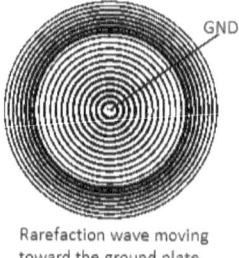

Rarefaction wave moving
toward the ground plate

Reactive Power Transmission pair

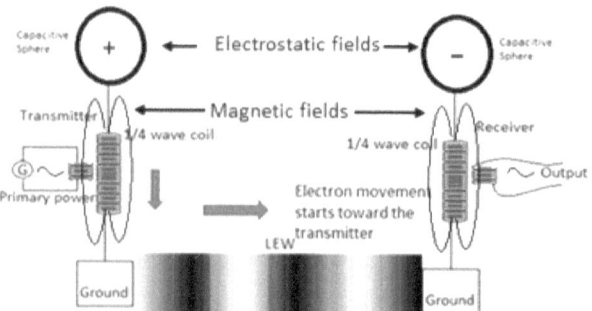

Figure 4.7-B Electron compression wave

3. The negative signal on the ground forces a compression wave toward the antipode.
4. The cycle begins its move toward 0 Volts: 90° - 180° slowly allowing a trickle of electrons to move up the coil toward the positively charged sphere
5. At 0 Volts point (180°) electrons begin to rush up through the ¼ wave coil toward the positive sphere
6. Depending on the frequency, the 360° cycle scenario will playout several times before the first wave reaches the receiver
7. The drawing of electrons into the sphere causes a rarefaction LEW at the ground plate
   The first LEWs have reached the receiver
8. The LEW compression wave will be the first LEW to encounter the resonant receiver's ground plate. The transmitter continually generates waves toward and from its antipode

Reactive Power Transmission pair

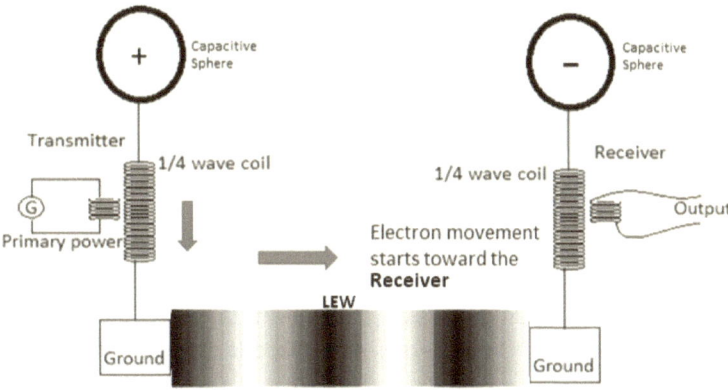

Figure 4.7-C Transmitter discharging

9. The receiver begins its negative charge cycle when the compression wave arrives. The LEW induces current and voltage into the ground plate of the receiver (and every other conductor in the earth too)

10. After the receiver sphere is charged negatively the rarefaction LEW arrives from the transmitter. The receiver sphere discharges its power into the ground plate at the perfect time in harmony with the rarefaction cycle.

Reactive Power Transmission pair

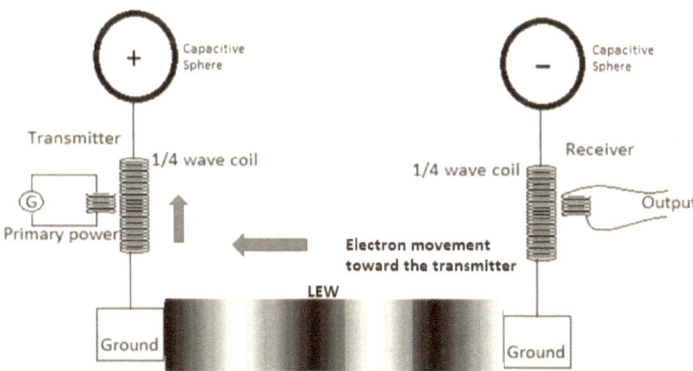

4.7-E Electrons begin to leave the receiver and move toward the transmitter

11. The LEWs from the receiver compression cycle move away from the ground plate and encounter new incoming LEWs from the transmitter at the same frequency

12. These LEWs are both negative and combine to double the power in a standing pattern. Fig. 4.7-E
13. The standing waves form in a series fashion as the receiver's LEWs encounter the transmitter's LEWs in succession. The number of waves in the standing pattern are dependent on distance from the transmitter and the resonant frequency.
14. When the standing wave forms power increases dramatically, doubling with each encounter. The power build-up is constrained by consumption at the receiver, the friction of resistance and reactance, or electrical spark discharge at a receiver or transmitter
15. The transmitter and receiver synchronize through the standing wave with the terminal capacitance and with opposite charges for optimum efficiency.
16. It is possible for harmonics to cause the transmitter and receiver to have the same polarity, this indicates there is an extra, odd standing wave because of imperfect tuning.
17. The system will function at a reduced power level
18. Power should be taken at the receiver secondary or sphere to prevent voltage overload

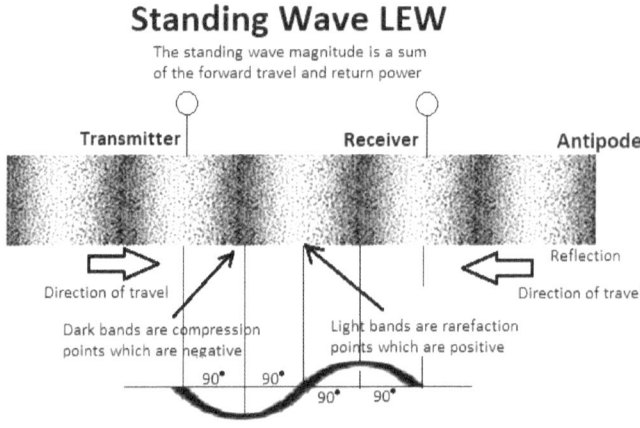

Figure 4.7-F The resonant LEW standing wave with the transmitter primary power input shown to clarify the AC power input and its relationship with the standing wave LEW

Figure 4.7-F compares the primary AC power waveform with the standing

LEW waveform. A sinewave equivalent of the LEW is measured with an oscilloscope at some point on the quarter-wave coil to earth ground. Vertical lines between the sinewave and the compression wave in 4.7-F show the sinewave in correspondence to LEW's pressure points. The sinewave follows the AC power supply generating them. To reiterate, there are two compression sequences per each 360° cycle, 180° apart. The resonant LEW always strengthens to the point where friction curtails it or the power is released in some way.

The peak power within a standing LEW is determined by power consumption at the receiver's secondary winding or capacitive terminal. In Figure 4.8 'Frame B', on the right, is the 'Top view' (*looking down from above*) picturing compression and rarefaction points in a standing LEW. The pressure points are moving in two directions but the pattern is standing still. The standing wave pattern results from LEWs moving between the transmitter and receiver. Standing LEW movement isn't visibly obvious.

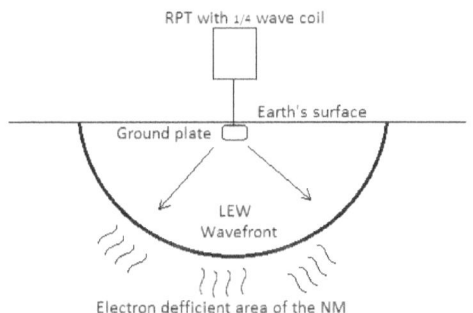

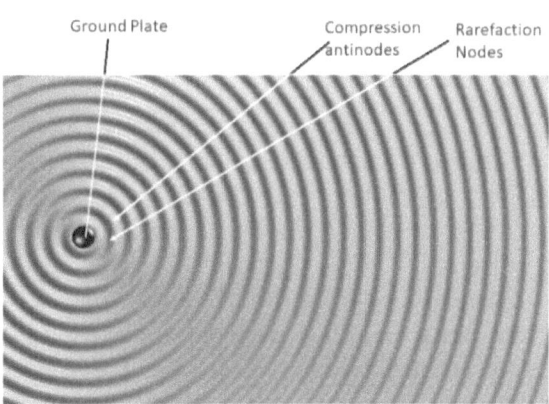

Frame A. **Figure 4.8 LEW wave-front at the ground interface** Frame B.

LEW movement

The LEW does not move in straight lines, but in a spherical and circular pattern as sound waves or [74]water surface waves do in one direction toward the antipode. Antinodes are always negative and nodes are positive. Positive antinodes do not exist in a resonant LEW. Nodes are zero LEW voltage points and are always positive compared with the antinode. A voltage drop exists between most points in the LEW. The exception is node to node or antinode to antinode. It is likely LEWs form [75]Chladni Patterns on the surface of the earth between resonant receivers and the transmitter. This would be expected because LEWS are light-speed mechanical (*sound waves*) moving in the earth's surface electron medium.

Extensive tests were made with LEWs in in an underwater setting. It was determined LEW moves through the water at all depths. They penetrate the bottom containment area of mud or sand and continues through the earth as a normal LEW. We examined various Tesla designs as well as our own customized ones. We did many tests with ferrite core transformers integrated into the Tesla design and stand-alone. I believe that this method of power transport is very versatile and will someday be the main energy transfer model.

LEW receivers

On our ferrite core Tesla transformer designs, tuning was done at the secondary side. We examined circuits with and without a quarter wave coil in series with the ferrite secondary. Other experiments were made with a C-3 (*a three-plate capacitor*). The C-3 presents a *reactive* receiving component coupled with a tuned ferrite transformer. The assemblies we built were earth-grounded, LEW reactive receivers, and altered from the Tesla design. Our conclusions were that ferrites offer more advantages in broadcast arrangements than in resonant ones.

The challenge with ferrites is their resonant cores. The engineering challenges in the ferrite transformer design are various. The variables of mutual induction, winding gain, parasitic capacitance, primary winding length, and the core's frequency range combine to complicate the design. The positive side is its compactness and high-power output.

We also explored various one-way broadcast types schemes. Each used a reactive design which is a single wire tuned circuit employing capacitive storage. We built synchronous transmitters and receivers with each half using a different frequency. We used these for one-way broadcast providing a quantity of usable power to the receiver. The LEW produced by this arrangement is a full three-hundred-sixty degree traveling wave with one rarefaction and one compression wave each cycle. Broadcast LEWs are lower in power, non-resonant, and are full cycle. Non-resonant LEW transmission is effectual for a limited range and can deliver power and communication. These types of transceivers are synchronous or asynchronous and used for broadcast-only application like radio or a Walkie-Talkie.

---

[74] https://www.youtube.com/watch?v=z63fJUROeN0
[75] https://en.wikipedia.org/wiki/Ernst_Chladni

Though a synchronous broadcast plan does not support a standing LEW it still uses a reactive design to receive and transmit. Our tests indicate that a higher frequency, above 500 kHz, is best for aerial power and communication but attenuation limits broadcast LEW range. Inductive reactance drains the LEW forcing voltage adjustments for greater reach. Synchronous broadcast can transport power and broadband. Various transport protocols, OFDM, FM, TDM and others are supported by LEW transport.

Underwater broadcast synchronous systems are possible. They can deliver power and broadband communication. It is possible to have submerged drones with 3D cameras, microphones and underwater acoustic speakers. Systems like these will allow researchers to go anywhere underwater and travel the depth of the sea in real time without a recharge. The streaming video, remote control, and sound can be experienced via 3D goggles offering an underwater exploration experience like aerial drones do now. Transmitter voltage, the broadcast voltage of the drone, and frequency selection set the range limits. Large craft like submarines would use a resonant LEW with a working range of the whole earth.

Underwater, light-speed LEWs can provide many services. Broadband Internet for underwater habitats, power delivery keeping batteries charged, or just plain old telephone service. Imagine how this ability can change our lives. As I write this book, a university in Boston has completed preliminary underwater tests successfully and ocean trials will begin in the summer.

Chapter 5: The Electron Plasma Field for Powering Automobiles and Aerial Systems

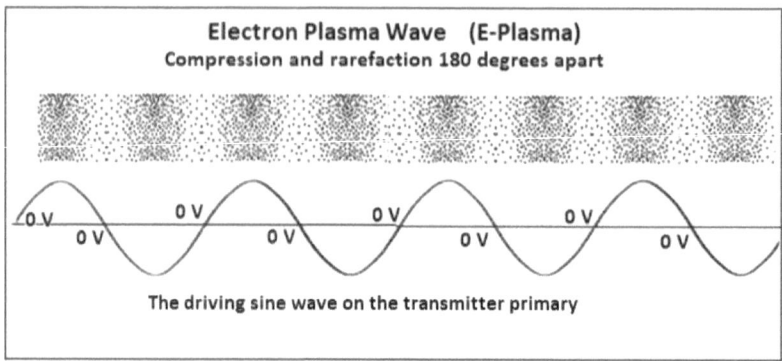

Figure 5.0 The Electron Plasma field (EP)

Surface LEWs support the Electron Plasma (*EP*) field. The EP field consists of vector fields from the positively charged upper atmosphere pointing to the negatively charged electrons in the earth. The field intensity varies in presence of the LEW. EP field variations produce power, not by magnetism, but by capacitive, electrostatic induction.

We proved the existence of [76] Electron plasma (EP) fields through experiments. This section on EP fields is of special interest to me because a promising facet of Tesla's wireless power transfer is revealed. Our tests cases yielded results that were scientifically noted many years ago. We used Maxwell's equations (*Gauss' law*) to express the EP field density which is principally a varying electrostatic charge. The equation describes the field strength and Ionospheric phenomenon. Electrical engineers, researchers, and science have dropped the ball in the study of the Tesla reactive power transfer method. The truth that longitudinal waves are created where power connects to ground has been missed altogether.

---

[76] *The information presented here has been authenticated through research by Wireless Power Technologies, USA. My own research and many of the ideas presented in this book are not available online or in textbooks, but are penned from daily logs from which the text in this book was derived. The various Tesla resources, written and online, are very helpful, but they do not address the electric field directly. Tesla himself presented certain logic and the mathematical proof in his writings, but does not mention the field that was to power his aerial systems. His world power plan included electric airplanes and dirigibles as well as wireless power transfer through the earth. Tesla fully knew about the strong electric field created by the earth currents, and he knew how to use them.*

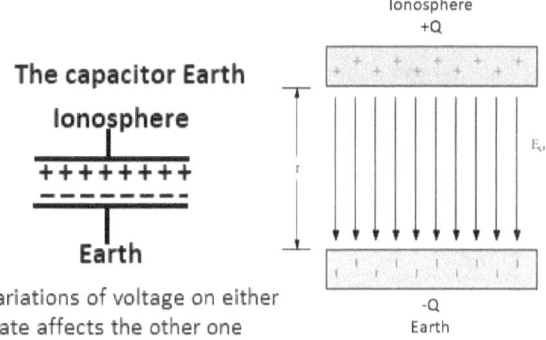

Figure 5.1 The Capacitor Earth
Earth is the (-) plate, the air is the dielectric, Ionosphere is the (+) plate

An Electron plasma (*EP*) field is not electromagnetic or a conveyed surface wave. It is an interesting phenomenon not typically studied in normal electrical work. Engineers studying electrostatics will be more familiar with the EP field than circuit designers or conventional, electrical engineers. The closest study relative to the electron plasma (*EP*) *field* is fulminology (*the study of lightning*). The electron plasma *wave*, is so called because it anchors to the LEW's concentrated waves of electrons. Electron Plasma is a positive vector field with lines of force terminating on the earth's surface electrons from the Ionosphere as shown in figure 5.1.

The effect of electric fields in circuit boards and wires is a rogue force. To circuit designers it the bane of sensitive low power circuit boards. This rogue does not easily lend itself to creative work nor is it easily confined. It lacks the power concentration of the magnetic field, but it causes added power consumption through parasitic capacitance. Electric fields are not as useful as magnetic ones, but they still pack a lot of power. The electric field is ever present and it is 'in charge' in obscure ways. Electrostatic polarity is responsible for current flow. Without electrostatic potential differences, batteries and power supplies could not work.

The capacitor is a common, passive electrical component. It stores electrostatic power, causes resonant oscillations with inductors, and has countless other electronic uses. Capacitors can hold an enormous amount of power that can be released in nanoseconds. Don't put your finger across the terminals of a charged capacitor or you will discover the power for yourself. Modern capacitors with improved dielectrics now replace rechargeable batteries in some small gadgets.

We all experience effects of Electrostatics. Have you ever slid across the car seat on a cold dry day and touched the metal of a car door handle? Or have you walked across a carpet and received a shock when your hand nears a door knob? This an effect of electrostatic energy. In these cases, you experienced the uncomfortable effect of becoming charged plate of several thousand volts.

Electrostatic energy discharges from a high potential (*positive*) to a low potential (*negative*) or sometimes called a sink. The word potential (*the root word, potent*) in electrical jargon connotes a difference between lower and higher voltage. The voltage difference between the plates define the capacitors potency. A car battery and a capacitor are similar with positive and negative terminals. The voltage of the terminals express [77]electrostatic potentials. In electrostatics, power flows from positive to negative referred to as conventional flow. Electrons move from negative to positive at a crawling pace of two to ten centimeters per-second. The real power is electrostatic, flowing at light speed from positive to negative against the bias of the electron flow. The LEW is a different case. Electrons do not flow in a LEW but they simply pass power by compression and rarefaction.

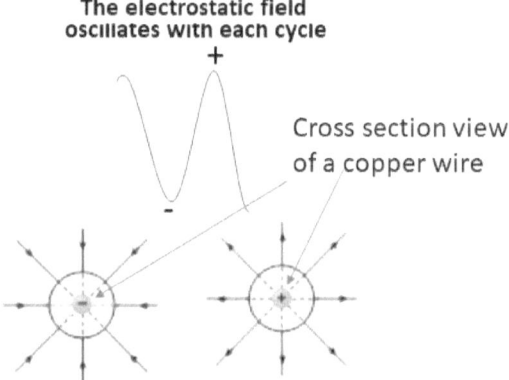

**Figure 5.2 Cross section view of the electric field of a wire:
AC power changes the polarity with every cycle**

When power flows through a conductor, a magnetic field spirals around it, and an electric field emanates at ninety degrees out from the surface of it, see figure 5.2. Normally, the magnetic field lines interfere with the electrostatic field lines and divert them. Electric fields in supply wire are ignored except in high frequency and radio work. At radio frequencies, electrostatics cause various problems, especially in circuit boards and integrated circuits. The parasitic capacitance between windings absorbs power in a Tesla coil: it modifies resonance, increases shocks, and causes sparks. Electrostatic forces are very powerful and must be given proper respect. Like Superman, they are 'able to leap tall buildings in a single bound', in the form of lightning. Electrostatic force lines can merge to become denser. They are a straight-line force but curve toward opposite charges or away from similar ones. In a Tesla coil the E-field linking effect (*merging of the 'E' field*) causes power to pass over the ¼ wave coil toward the terminal or ground depending on the potential of the wave in the coil. This changes the frequency of the resonance and absorbs energy intended to charge the capacitive terminal or the ground plate.

---

[77] https://physics.stackexchange.com/questions/135779/how-does-a-battery-work

The Ionosphere:

The Ionosphere is the positive plate of this 'supersized' capacitor we know as earth. [78]The Ionosphere is under continual bombardment by positive ions (*charged particles*) from the sun and cosmic particles from outside our solar system. These Ions pass their charges off as they collide with particles in the (*conductive*) Ionosphere. The charge increases giving the Ionosphere a high positive charge.

A positively [79]Charged capacitor plate emit lines of force attracting negatively charged electrons to the surface of the opposite plate. This is the case with the earth, see figures 5.3 and 5.4. The electrons become a sink for the positive lines of force and they become electrostatically linked to the Ionosphere. This coupling is an electrostatic connection between the two plates causing equal but opposite voltage potential. In any capacitor, a positive charge placed on one capacitor plate causes an equal and opposite effect on the other. This is called capacitive coupling.

---

[78] https://ntrs.nasa.gov/archive/nasa/casi.ntrs.nasa.gov/19650014232.pdf section III, electric fields & Ionosphere
[79] https://www.build-electronic-circuits.com/how-does-a-capacitor-work/

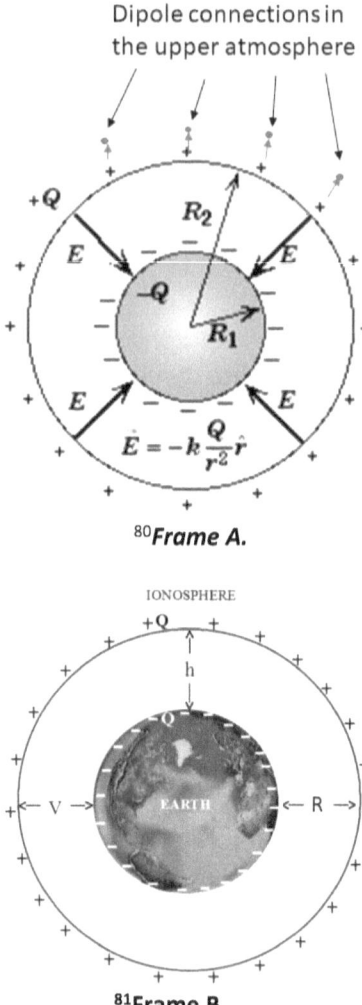

Figure 5.3 The earth is a charged capacitor

The positively charged Ionosphere draws the earth's interstitial electrons to the surface, figure 5.3 frames A&B. A positive '+' vector field from the Ionosphere links to them creating a [82]dipole, see 5.4-A. The figures 5.4-B and 5.5 below show how the lines of force connect between earth charges. A simple example of charge flow is the kitchen sink model with the water running and the sink drain open. The water flowing from the faucet represents the Ionosphere's source vector, and the

---

[80] https://www.sciencedirect.com/science/article/pii/S1364682617303711
[81] http://www.abovetopsecret.com/forum/thread780157/pg1
[82] A *physical dipole* consists of two equal and opposite point charges: in the literal sense, two poles. Its field at large distances (i.e., distances large in comparison to the separation of the poles) depends almost entirely on the dipole moment.

surface electrons are the drain. All charged particles (*figure 5.5*) are either a source (+) or a sink (-) and when connected they are dipoles. Charge flows constantly from positive to negative, from the Ionosphere to the earth in this case.

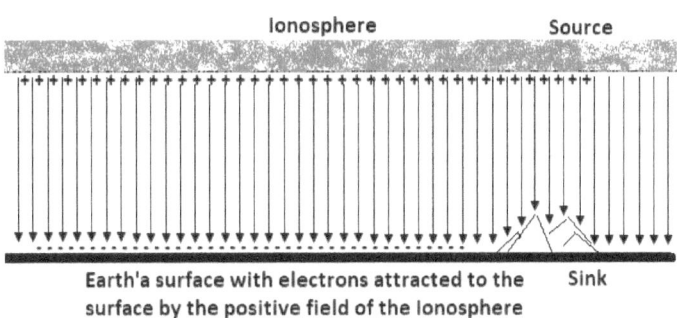

Figure 5.4-A The lines of force from the Ionosphere to the earth

Ionospheric ions (+) and anions (-) endure continual agitation by sunlight, rotational forces, and particle bombardment. A NASA paper from 1965 [83]developed a technical description of this electrical activity. The study of rocket probe data revealed these facts.   Sunlight causes a chemical reaction in the Ionosphere releasing electrons called a *diurnal chemical reaction*. Atmospheric Ions and the new, chemically released free electrons in the Ionosphere become connected forming electric dipoles until evening. This has a direct effect on earth's daylight, surface electron density.   When the Ions switch their connections to nearer electrons in the upper atmosphere the lines connected to the earth are released resulting in a voltage drop (*a charge difference*) between the earth and the Ionosphere.

The dipole switching event is a slow transition of Ionospheric lines of force from earth's electrons to the new diurnal atmospheric electrons. When the surface electrons are no longer constrained by force, they migrate below the surface into an interstitial domain where they will stay until evening. The reduced charge density is responsible for the voltage drop between the earth and Ionosphere.

Diurnal [84]Flux density is reduced by one-tenth (250,000 *V down to 225,000 V*) of the nocturnal ionospheric density.   The ten percent daytime loss in electron

---

[83] https://ntrs.nasa.gov/archive/nasa/casi.ntrs.nasa.gov/19650014232.pdf (Adiabatic motion of Auroral Particles in a Model of the Electric and Magnetic Fields Surrounding the Earth by Harold E. Taylor and Edward W. Hones, Jr.)
[84] http://www.dtic.mil/dtic/tr/fulltext/u2/a088879.pdf

density conveys directly to an equal loss in LEW voltage strength. To support power and communication links the primary voltage must be increased. Turn-up testing and commissioning usually occurs in the daylight hours. Consequently, this balances the LEW for diurnal losses. Nighttime becomes a bonus for RPT (*Reactive Power Transport*) wireless power systems, more power is available for charging batteries on mobile units taken off-grid during high demand time.

A side effect of the nocturnal and diurnal electron field density variation is a current flow from movement of electrons from the surface to the lower parts of the earth. The upward and downward electron flow creates electric and magnetic fields. This is an interesting event with unknown ramifications. What shape are magnetic fields caused by electrons migrating from surface toward the center and back out again later? How about the electric fields? These are interesting questions with no answer.

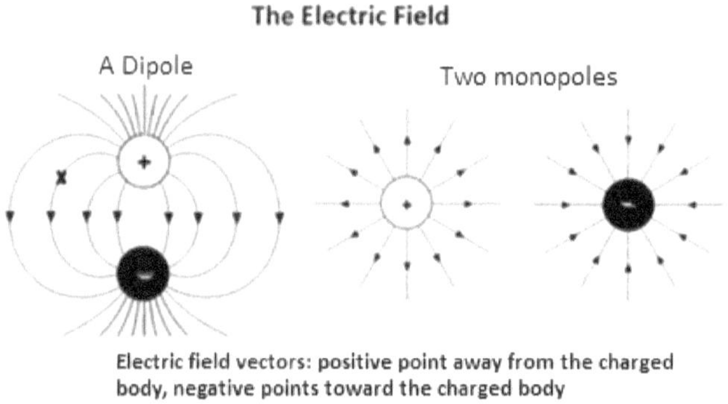

The Electric Field

Electric field vectors: positive point away from the charged body, negative points toward the charged body

Figure 5.4-B Electric dipole with its constituent charges

Seasonal changes can degrade the efficiency of a Tesla RPT network. The extra hours of daylight during summer in either hemisphere reduces electron density in the medium for longer periods of the day. The lower electron density causes 'sponginess' in the medium and electrical resistance for the LEW. The electron density reduces the LEW's potency and calls for a higher voltage at the transmitter to preserve the standing wave. Failure to increase power can result in resonant signal collapse (*RSC*) and loss of the standing wave. A voltage increase in the primary power is necessary at the transmitter or the condition will continue and cause recurrent failure until the electron density increases in the evening.

The extra energy needed to preserve the LEW in daylight is reactive power and it is returned to the source each cycle. Higher voltage is always necessary to preserve the LEW and the EP field during daylight hours. Deep, multiple pole grounding at the transmitter and receiver will limit the effects of sunlight loss on LEW transport, but the solar effect on dipole field density between the Ionosphere and the earth still exists. This limits EP flux density (*EP field density*) between the earth and the Ionosphere *even with increased* power in the transmitter.

Aerial apparatuses use the EP (*electron plasma*) field for power. The standing wave LEW concentrates the EP field at compression points and dilutes them at rarefaction points. The electrostatic flux is a non-thermal (*cold plasma*) field existing between the sky and earth. It is a mixture of various ions and other charged particles agitated by the LEW movement of many frequencies. In figure 5.5 the EP field modulated by a LEW is referred to as an EP wave. The wave provides aerial power for cars and drones.

An aerial receiver obtains power without an earth connection. Capacitive coupling in an agitated EP field provides the means of power transfer. Properly designed aerial systems can receive enough power to fully charge batteries while in operational service. Proper scaling of the aerial receiver/s and transmitter/s is essential to receive acceptable power amplitude for transmitting data. The LEW transfers the power to the aerial receiver supporting power demand for cars, buses, drones, or other devices.

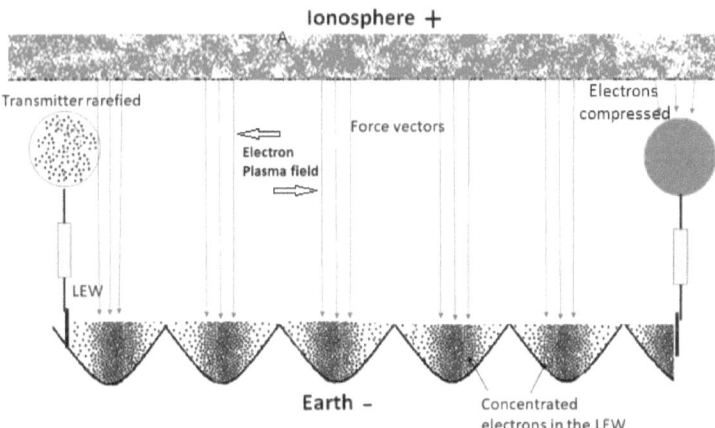

Figure 5.5 Dipole connections between Ionosphere and earth electrons

EP fields are similar to the electrostatic field charge sensed by some people during an electrical storm near the point of a lightning strike. A concentrated EP field (*near an RPT transmitter or receiver*) presents a sensation of tingling and causes electrostatic effects like those created by a Van DeGraff generator of hair standing on the head and arms.

High voltage EP fields cause florescent and neon bulbs to glow near charged Tesla coils or other EP field sources. EP fields following a LEW are too weak to cause any drastic effects. Tesla mentioned in his Colorado Springs Notes of 1900, and I paraphrase, "The moths in the air glowed green, the horses' metal shoes had sparks jumping to them from the earth when they were near the transmitter, and insects near the coil were immediately killed". He built the first a modern-day bug zapper.

An example of a natural EP field effect is 'ball-lightning' occasionally witnessed during electrical storms. Flaming is due to electrostatic charges and is like glowing of fluorescent or neon tubes placed near a Tesla coil or other high electrostatic source. A rarified air bubble forms during a lightning strike. The strike chemically combines nitrogen and oxygen making solid nitrates in a volume of air leaving a rarefied gas. The rarefied gas glows in the presence of an EP field. The 'ball's' volume is expressed by the heat of the rarefied air. The spectacle lasts only a few seconds because of the slight vacuum and the wispy nature of the spinning gas, and/or the EP field charge duration.

A quick summary: In figures 5.5 and 5.1 above, the Ionosphere is the positive plate, the earth is the negative plate, and the atmosphere is the dielectric. The density of electrons at LEW compression points modulates with the standing wave LEW. The [85]modulation converts the EP field (*vector force lines*) to an EP wave.

This is good news for users of wireless. Aerial drones, planes, cars, and cell phones can be used anywhere on or above the earth, the EP wave range extends to the Ionosphere. Power is picked off through resonance or through a timed interrupter.

To get real power, place a reactive receiver anywhere in the atmosphere (*dielectric*) where there are EP waves. A simple analogue receiver consists of two or more plates on each side of a coil. The more advanced receiver is a switched transistor or MOSFET (*figure 5.6*) variety. To achieve maximum power the receiver switches in synchronization with the changing flux density of the LEW. This works on the same principle as Tesla's patent [86]685953. The receiver works aerially or with two plates planted in the earth separated by one LEW's wavelength.

---

[85] EP field becomes an EP wave when the field is modulated by a Longitudinal Electron Wave (LEW)
[86] https://teslauniverse.com/nikola-tesla/patents/us-patent-685953-method-intensifying-and-utilizing-effects-transmitted-through

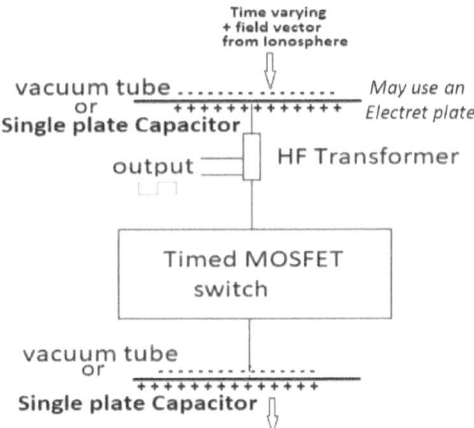

Figure 5.6 A simple example of a timed aerial receiver for drones and cars

EP field strength:

We proved the existence of EP fields and their inherent power with a receiver built to resonate with a LEW from our transmitter. We did tests to ascertain the efficiency (*percentage*) of the transferred power to the aerial receiver. We did not want to commit time and money on receiver development until we knew this answer. We partially answered the question in an experiment we completed in mid-2017 at Wireless Power Technologies' lab.

Later in this chapter I mention Tesla's *PierceArrow* experiment. I had interest in this event because it speaks directly to the reality of wireless power for aerial use. Part of the article reveals that Tesla had a wireless power 75 kW gasoline-powered station in Canada. This fact is mentioned in a [87]Peter Savo account. The account mentioned Tesla used a vacuum tube receiver that received enough power to propel an eighty-horse power electric motor placed in a Pierce Arrow car at ninety miles per hour. If the account is true, then this is proof the LEW and EP field are powerful enough to run a vehicle. We have not yet built a transmitter to the scale required to do this.

We made a simple Tesla reactive receiver (*figure 5.7 below*) with capacitor plates on either side, top and bottom. The top one was a nine-inch diameter, titanium-dioxide coated sphere, and the bottom plate was a twelve-inch square copper sheet. We raised the receiver from 5 to 25 feet up on a fiberglass mast. We managed vertical adjustment with a dry cotton cord and a pully at the top of the

---

[87] http://www.tfcbooks.com/teslafaq/q&a_016.htm

mast. We connected to the aerial receiver secondary winding with a twenty-five-foot COAX cable to our test gear. The test gear was an oscilloscope and a small carbon resistor commonly used as a load for a testing reference.

The aerial receiver had a three-hundred-fifty Kilohertz quarter wave primary winding and a 10:1, low voltage output. An HP 214B high voltage pulse generator was directly connected to the ground plate and used as the transmitter. The 214B connected to a 110VAC isolation transformer with open grounds between input and output. The pulsing LEW and EP wave was smoothed by the receiver's quarter wave air transformer coils into a skewed sine wave. The aerial receiver provided the same output voltage as the normal, grounded, resonant receiver. This confirmed the EP wave is an efficient conveyer of power.

An EP field wave driving an aerial receiver

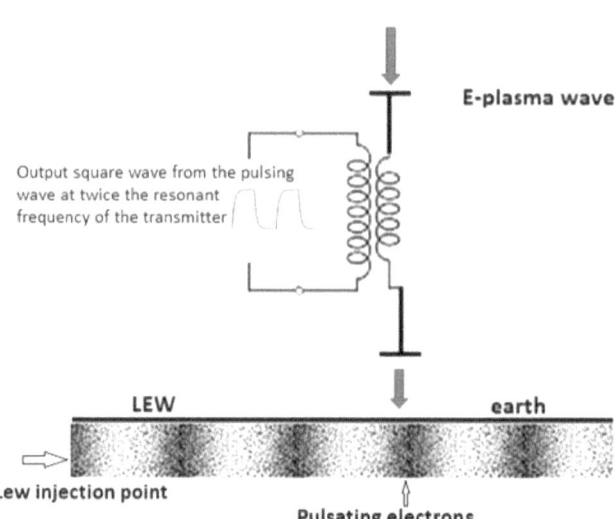

Figure 5.7 A passive EP field aerial receiver

The peak voltage of the LEW compression points defines the EP field strength. The voltage (*EP field strength*) of the LEW varies according to the standing wave power stored in the LEW.

The EP field density between the Ionosphere and the earth strengthens with higher LEW voltage. We use Gauss's formula for charge density to calculate the EP density. In Maxwell's equations, there is a subset of equations describing electric and magnetic fields. The Gauss part of the equation defines the electric flux density D by using electric charge density 'ρ'.

Gauss defines flux density by the formula $\nabla \cdot D = \rho_V$. The diverging ($\nabla \cdot$ *divergence operator*) flux density (D) is directly proportional to the electric charge density. The divergence source is the positively charged Ionosphere with vectors pointing toward the electron sink (*the earth*). This definition describes a capacitor, which we know the earth is.

The energy (U) (*in Joules*) of a capacitor is expressed by the charge Q^2 (in Coulombs) divided by twice the capacitance (C), stated $U = Q^2 / (2C)$. The earth has roughly [88]500,000 coulombs of charge and the capacitance ≈ 711 UFD. The term 'U' is solved to be 175,808,720,112,517 Joules, or in more familiar terms 175,808.7 Megawatts per day. The largest nuclear power plant in the United States is Palo Verde, it produces 3,937 Megawatts per day. The continual dielectric current leak between the Ionosphere and the earth causes a steady electrical discharge.

More facts and equations:

1. There are lines of force between the Ionosphere and the earth surface because the earth is a super-large capacitor with a capacitance of about 711µFd; there are lines of force between the plates of a capacitor, within the dielectric, pointing from positive to negative.

2. According to Gauss's law, capacitors contain a force, **F**, which exists between charges on the plates defined by the formula $F = q_1 q_2/4\pi\varepsilon_o R^2$: R is 48,280 meters, the distance between the Ionosphere and the earth, the q1 charge is the Ionospheric positive charge, and q2 is the earth's negative charge which is $\approx$ [89]$4*10^5$ Coulomb. q2 is the sink on which the ionospheric electric field lines from q1, terminate, and epsilon naught, ε_0 is $8.9*10^{-12}$.

The calculation: using the Force equation then substituting q_1 and q_2 with $4*10^5$, and 42280 Km as R yields $F = ((4*10^5) * (4* 10^5))/4\pi*(8.9*10^{-12}) * 42280^2 \approx 2*10^8$ Newtons. This force exerted between the Ionosphere and the earth's surface contributes substantially to atmospheric pressure. The force pulling downward on the Ionosphere generates spherical containment for the Troposphere, Stratosphere, Mesosphere, and part of the Thermosphere below it.

3. The E (electric) field over the surface of the NM (*earth*) is defined as $|E| = q/4\pi\varepsilon_o R^2$, substituting yields $|E| = 4*10^5/4\pi*(8.9*10^{-12}) * 3959^2 = 2.28*10^8$ Coulombs distributed over the surface of the whole earth. Reducing this number to square meters yields $2.28*10^8 c /1.97*10^8$ $M^2 = 1.157$ C/M^2 (*Coulombs per square meter*). The number of electrons in one square meter

[88] https://www.physicsforums.com/threads/net-charge-of-the-surface-of-the-earth.771471/
[89] http://physics.oregonstate.edu/~mcintyre/COURSES/ph431_F12/examples/EarthCharge.pdf

can be calculated: *6.24*10[18]* electrons/Coulomb * 1.157Coulomb/M^2 ≈ 7.22 *10^{18} free electrons in every square meter.

Positive charges on the Ionosphere mobilizes earth electrons. Every surface electron has at least one force line connected to the Ionosphere. When the surface electrons become agitated at a high frequency by a LEW, they modulate the EP field allowing the transfer of power to aerial devices.

The previous drill describes in mathematical terms the strong force of electric fields between the earth and the upper atmosphere. Diverging fields (+) from the Ionosphere to the sink of individual electrons in the earth (-) and other charged particles that forms the EP field. The resulting field is a nonthermal, 'E' (*Electric*) field also known technically as [90]cold plasma.

Review: The earth E field has a field strength of 113.5 Coulombs per square Km of the earth's surface. The LEW oscillations agitate the electrons and the flux lines (*[91]D and E field*) connecting to them. The EP wave induces power in an aerial receiver through capacitive coupling by causing a potential difference of voltage between the plates of an aerial receiver. The potential difference (*figure 5.8*) induces a current flow in the transformer's primary winding. The secondary winding delivers the power used for propulsion, communication, or other purposes.

A simple analogy of this event is capacitive coupling. A signal impressed on one capacitor plate changes the opposite plate 180° degrees out of phase. The connecting lines of force change plate voltage through the dielectric. A positive signal on one plate causes electrons to bunch on the other and vice versa. The atmosphere is the dielectric of the earth capacitor. When a receiver is within the dielectric, changing flux lines charge each plate oppositely. The potential difference moves charges (*not electrons*) from positive to negative through the receiver transformer to the opposite plate. The moving charges are electric power. Charges flow from positive to negative, electrons flow from negative to positive. Charges flow between capacitor plates and not electrons.

Magnetism and transformers:

Magnetism plays two roles in Tesla RPT oscillators. First for transferring power by transformer action and second for energy storage in a magnetic field. The magnetic field releases in harmony with the driving primary signal enlarging power output. At the ground plate, high voltage electrons enter the earth and detonate a compression wave (LEW) in the NM. The high voltage electrons from the transceiver repel the lower potential earth electrons in an explosion of power creating the EP

[90] https://www.sciencedirect.com/topics/chemistry/cold-plasma
[91] https://en.wikipedia.org/wiki/Electric_displacement_field

field fluctuation. The EP field is very powerful around a transceiver, especially near electrostatic components.

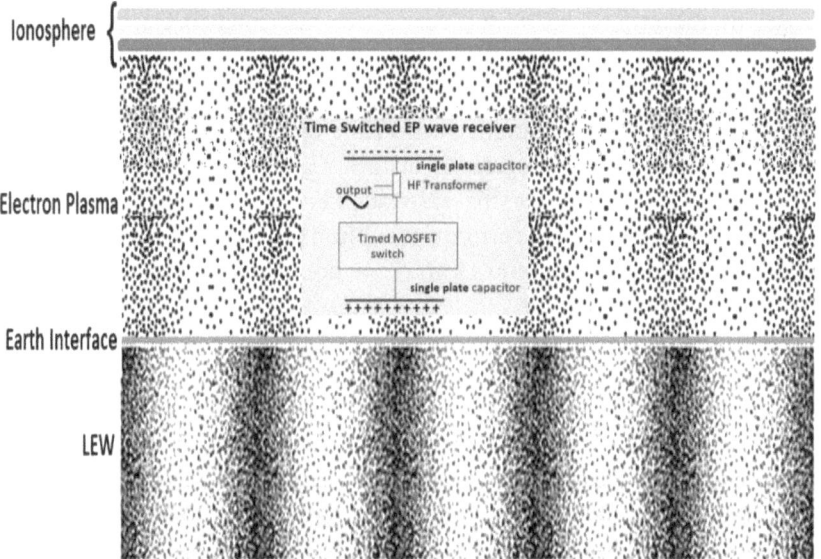

Figure 5.8 An active receiver utilizing the EP field for power

EP wave receivers:

There are two types of EP receivers, passive and dynamic. Figure 5.8 is an active, time-switched MOSFET receiver, and the passive version is shown in figure 5.7. The passive version has a quarter wave coil bounded by two single plate capacitors; it is usually too large for aerial drones, but well suited for an automobile. Supercooling the quarter wave coil boosts performance of a transceiver.

The dynamic receiver is compact and needs a power source to work. At startup, synchronization with the EP wave begins. After the receiver locks to the EP wave, a surplus of power is generated. A proper design provides power for MOSFET switching and battery charging. There are many variations to this scheme.

EP field limitations

EP fields do not penetrate underground or under water. They are an atmospheric phenomenon therefore normal mobile receivers will not work submerged. In a subterranean or submerged environment mobile receiver may use a positive [92]electret connected to an internal, terminal plate as one side of the receiver. The electret creates a positive field which delivers a full 360° oscillation from the negative-pulsing, resonant LEW. Subsea receivers in a non-conductive hull vessel can use a copper plate on the outer hull as one connection and an electret plate inside the submersible. A submerged receiver within a metal enclosure can use the outside surface as the earth-ground plate and a dry, insulated plate or sphere inside as the other. Inside the conductive hull submersible, a standard vacuum tube or spherical terminal is used.

Stationary, subsea or subterranean receivers use two conductive plates in placed in the NM or a grounded, regular oscillator. If ground plates are used, they should be placed ¼ wave apart in the resonant LEW. They must be in direct alignment with the LEW for best performance. The LEW can penetrate 21 miles under the earth (*calculated using Tesla's c*π/2*) and ocean and transfer power and a light-speed broadband data link.

Automobiles and aerial drones need special attention in their electrical design. Any receiver without a ground connection is considered aerial. As stated before, they connect to the LEW through an EP field by capacitive induction. Resonant [93]RPT receivers continuously charge batteries and provide direct power to a consumer. Battery power is the primary power supply in mobile receivers and the RPT serves as the mobile charging station. Mobile receivers experience narrow dead spots called Null Points making batteries important to sustain steady power. Charging resumes at a normal rate beyond the null point.

---

[92] https://en.wikipedia.org/wiki/Electret , https://www.britannica.com/science/electret
[93] Reactive Power Transfer

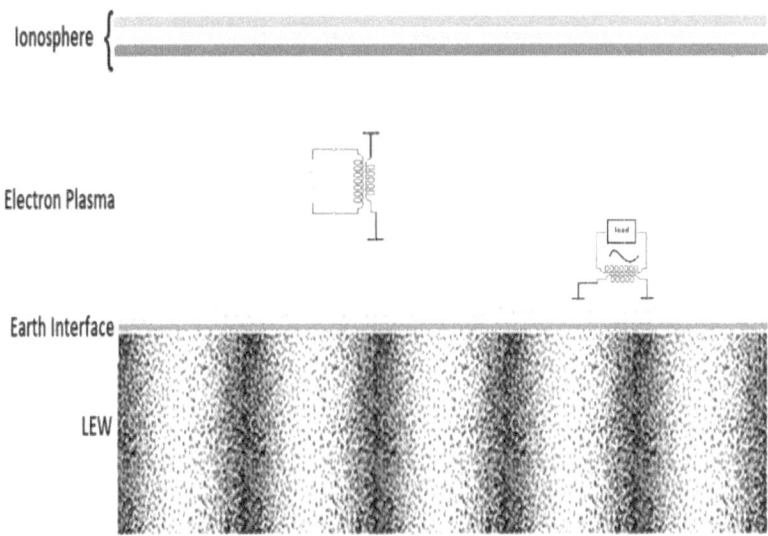

Figure 5.8 A passive aerial receiving device

Standard electrical meters cannot measure EP field density of a LEW, neither have proper tools been made except for electrostatic field readers. A time-switched meter must synchronize with the LEW; an electronic version of Tesla's patent "[94]Utilizing the effects transmitted through the natural medium". A MOSFET switching array and a ferrite transformer nicely duplicated the mechanics of Tesla's timed receiver. Another method is to use a calibrated, resonant aerial receiver with a meter or scope across the secondary.

Our bench calibration was done by measuring the transmitter's secondary voltage and the current through the ground wire. These measurements provided a scale for comparing the receiver and transmitter power. The gadget has its limits, but it provides an applicable standard for comparing various receivers. The ferrite transformer core was selected to match the frequency range of the transmitter. Voltage taps were wound on the secondary to measure weak and strong signals. A carbon resistor load and a tuning capacitor were used on the secondary. This receiver arrangement works without a timed switch. The tuning capacitor is used to tune to the LEW's frequency.

The Pierce Arrow story

One fascinating Tesla EP field story comes to mind that has circulated on the Web for a while. The story says Tesla had a Pierce Arrow car retrofitted with an

---

[94] https://teslauniverse.com/nikola-tesla/patents/us-patent-685955-apparatus-utilizing-effects-transmitted-distance-receiving and https://teslauniverse.com/nikola-tesla/patents/us-patent-685953-method-intensifying-and-utilizing-effects-transmitted-through

electric motor, and used his wireless power system to power it. The original article appeared in the [95]*New York Daily News*, April 2, 1934 titled **"Tesla's Wireless Power Dream Nears Reality"**. It mentions the planned *"test run of a motor car for 30 or 40 miles over a stretch of Atchison, Topeka and Santa Fe railway track running from Boise City, Oklahoma to Farley, N. M. using wireless transmission of electrical energy to power the vehicle. The equipment was assembled by 'two Californians' and is described as including a high-powered radio transmitter with big coils and short antenna."*

The article goes on to confirm what I have read in several places, but hasn't been proven until this wireless car episode, that Tesla had a Colorado Springs style wireless power plant in Canada. The story goes on to say the Tesla transmitter was near Sandford, Quebec, Canada in some woods, and powered by a 75-kW gasoline generator. I Googled the distance from Boise City to Sanford Quebec, it is 2,000 miles. This power traveled through the earth 2000 miles and provided power for the car; not too surprising when you consider the earth has .0055 ohms from point to its antipode and Tesla liked to use 9300 Hz.

The article also says, "The Pierce Arrow test was demonstrated at top speeds of 90 mph, and the car never needed a charge; the power was maintained through the wireless. Tesla himself built the receiver on site with parts from a radio store which included vacuum tubes and other electronic components." Apparently, the rails the car was riding on provided the earth ground through the wheels. Tesla did this test in 1933 or 34, just ten years from the time when he said, in 1922, that he did not expect to see automobiles being wireless anytime soon because of technical challenges. It appears Tesla discovered how to convert his mechanical interrupter (*patent 787,412*) into the active EP field receiver built with vacuum tubes to provide his power. There was no mention of an oscillator receiver being installed. Information about the components he used are on the [96]hutchisoneffect Web site.

1) [97]12 Vacuum Tubes (70L7-GT rectifier beam power tubes or possibly 21-A)
2) Wires
3) Assorted Resistors
4) 1/4" diameter rods 3" in length

If you are curious, there are many 'hits' returned on Web searches about this event, but little information is returned. One must draw some conclusion with the information available. I think the car's battery was used to operate the receiver's vacuum tubes. Tesla probably used the diode tubes to rectify the AC power from the LEW. The triodes provided the power to the coils of the 75 HP motor, and kept

[95] http://www.tfcbooks.com/teslafaq/q&a_016.htm (excerpt from the book)
[96] http://www.hutchisoneffect.com/
[97] http://www.hutchisoneffect.com/Tesla%20Pierce-Arrow.php

the car battery charged. Vacuum tubes are high power devices. They are the driving force of 10 kW to 100 MW RF power devices. Solid state devices are improving in power density but vacuum tubes rule the high voltage arena.

As a side note, Studebaker bought the Pierce Arrow company in 1928. Studebaker went bankrupt during the Great Depression in 1933. Tesla would have never had success with the Studebaker owned Pierce Arrow because of the financial condition of the company. Even with a booming economy and strong interest in Studebaker products, Tesla would have been stopped in his tracks with his Pierce Arrow success by the oil companies of the time. They would perceive the electric car as an enemy to their financial 'bottom line'.

Chapter 6 - LEW (Longitudinal Electron Wave) Receivers

In chapter four, I discussed the LEW in detail, and there were diagrams of receivers and transmitters used to receive and transfer power through the earth. This chapter will reveal the electrical mechanics of various receivers and how they interface with the LEW. I will reiterate some of the points mentioned in previous chapters for clarity.

In the Tesla RPT (*Reactive Power Transfer*) plan, resonance is established with a prime receiver or several receivers. There are two important topographies: the full wave LEW, and the magnifying power of the resonance standing, half-wave LEW. The full wave LEW lacks the voltage and power to reach the antipode or a resonant receiver. Whereas any wave full-wave LEW reaching the antipode will reflect and form a standing half wave LEW pattern of twice the power.

Resonant receivers mirror antipode reflection. They establish standing waves with a transmitter as the antipode does. A LEW echoed by a resonant receiver grows in strength quicker through constructive interference. The wave may transverse the whole globe, depending on various circumstances. The LEW will gain the greatest potential between resonantly coupled transceivers. LEWs moving in various directions other than toward the receiver are weaker and skewed by harmonics. The EP wave and the LEW power transport capacity increases if two power transmitters run one-hundred-eighty degrees out of phase on the same resonant frequency (*near zero ERL*). Transceivers must be geographically placed to deliver the preferred standing wave.

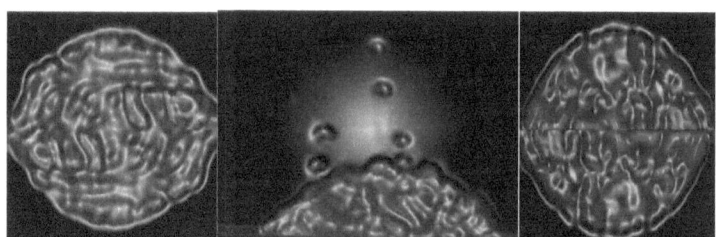

**Figure 6.1 Water drops placed on a speaker cone in zero gravity
on the ISS and influenced by low frequency sound waves**

Figure 6.1 illustrates a water drop experiment done on the ISS (*International Space Station*) in a low gravity environment. The center frame is a low frequency soundwave induced on a droplet attached to a small speaker. In the left and right frames, the half sphere pictures were placed bottom to bottom to form a full sphere. The left and right water drops were disturbed by different frequencies. If LEWs were visible from space, they would look similar. The center frame displays a wave with enough magnitude to spurt the medium into space. [98]Youtube.com has this experiment in action.

It is logical that a powerful LEW created by lightning (*or a powerful Tesla RPT*) creates an EP field with enough potential to release electrons and charged particles at the antipode. The electrostatic effects alone can cause a 'glowing ball' from ionized or [99]noble atmospheric gases. I have witnessed this effect near a charged Tesla coil with neon and fluorescent tubes. Ionization glow stems from positive EP field vectors locked-on to the electrons (*from the medium*) as they spew into the atmosphere. The center frame in figure 6.1 shows the effect where water droplets are ejected from the water bead. These ejected electrons provide the electrostatic charge for an atmospheric glow.

The Figure 6.2 represents a single transmitter with an antipode and no resonant receiver with a simple standing wave. A resonant receiver complicates the standing wave pattern and creates certain constructive and destructive wave points.

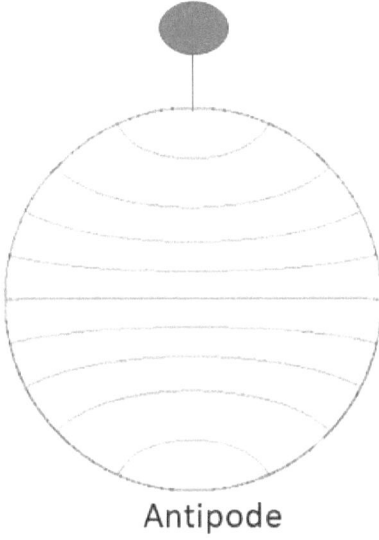

Antipode

Figure 6.2 Single standing wave between transmitter and its antipode

Destructive [100]resonance is not normal in nature, other than earthquakes, or weather events. Most of us have experienced the power of resonance in some way. The squealing speakerphone, sound system feed-back, a singer breaking glass, or other such effects are results of resonance, and compression sound waves are responsible. The RPT transceivers create resonant LEWs in a similar way to a playground swing. A natural, almost automatic swing with the legs 'pumps' the

---

[98] https://www.youtube.com/watch?v=B1u3SYmWbqo
[99] https://en.wikipedia.org/wiki/Noble_gas
[100] https://www.merriam-webster.com/dictionary/resonance

energy in. If the pumping action is done perfectly, height and speed are increased. Similarly, the Tesla RPT transmitter and receiver work in perfect harmony to add more power in the LEW with each stroke. The power placed into the LEW strengthens it and provides dependable output at the receiver secondary or at the terminal. But if power consumed exceeds power placed in at the transmitter then resonant signal collapse (*RSC*) will occur.

Mechanical or electrical resonance is difficult to control. Tesla told a story about an occasion where he caused turmoil in his lab building. He created a resonant mechanical machine in the [101]1890s he called the Oscillator. Tesla's intent was to make a precision electrical AC generator. He clamped the oscillator to a building girder for a test run and tuned it to the resonant frequency of the building girder. Soon the building began to violently shake and rumble like an earthquake. Tesla immediately stopped the oscillator, but it was too late to prevent pandemonium. Police came and evacuated the building. He mentioned nothing of this until twenty years later.

Nikola Tesla knew about the destructive power of resonance, and he had plans in place to control the effect in his Wardenclyff resonator. He had a very bad experience at his Colorado Springs plant. Resonant buildup in the LEW fed backwards, overdriving the quarter wave secondary. Transformer action induced power backward into the Westinghouse transformer secondary and fed high voltage and current down the line to the power plant burning the generators out. It happened so quickly that Tesla couldn't stop it.

Normally, resonance supports the LEW's power. However, during times of low power consumption LEW power will run-away if not kept in check. There are two ways to control resonant build-up in an RPT resonator. Shutting off or lowering power input on the primary of the transmitter is the preferred method. But during emergencies opening the ground connection on the transmitter immediately shuts the RPT down. Reducing power at the transmitter will allow the operator to preserve service and avoid complete shutdowns. Charging backup battery banks is a good plan for redirecting extra power to prevent resonant overload and offer a power conservation measure.

In 1901, Tesla was beginning construction on his dream machine, the Wardenclyff power station, figure 6.3, on Long Island, NY. Tesla had a plan to reduce electric friction (*resistance*) in the transmitter's primary, secondary, and the power feeds from the generator house located two hundred feet away. Tesla defined what he termed the magnifying factor, covered in an earlier chapter, for his transmitter. Resistance in the transmitter's coils lowers power capacity of the

---

[101] https://en.wikipedia.org/wiki/Tesla%27s_oscillator and
https://patents.google.com/patent/US685012A/en

oscillator. Tesla defined the magnifying factor as M = L/R (*I use Q_m – magnification quality here*) which is related to the Q factor of normal receivers and transmitters.

Tesla was experimenting with [102]superconductivity in copper coils and such, and he held patents in cryogenic conductors and insulation methods. The *reactance* of the coil adds to its resistance, Q_m = L/X_L+R. Resistance (R) is consistent across frequencies. Inductive reactance varies directly with frequency by X_L = 2πFL: where X_L is the inductive reactance (*AC varying resistance*), F is frequency, and L is inductance. Reactance (X_L) is not influenced by cryogenic freezing. But by reducing R to near zero with cooling increases the magnifying factor Q_m to an extraordinary level. Tesla's plan of using low resistance and low frequency increased the power and range of his transmitter. An interesting side effect of zero resistance, is no voltage drop exists between windings of the coil and therefore no parasitic capacitance. Tesla killed two giants with one stone.

Let's look closer at the figures for the Tesla Q_m factor with a cryogenic resistance versus a normal one. Tesla had several interchangeable secondaries for his transmitter at his lab. The secondary coil on page 206 of his [103]C.S. notes was most often used. The secondary coil was wound with number 10 wire, eighteen turns on a 49-foot diameter form. The coil had a quarter wave length of 2,770.885 feet *(.5248 mi.)*. The inductance (L) was 2.06 H, the resistance (R) was 2.804 ohms, and the resonant frequency was 88,607 Hz, a high earth harmonic (*88607/15*). The normal inductive reactance (XL) was 217Ω and the magnifying factor of this coil was Q_m = L/R = 2.06/2.804 ≈ 0.7347.

Cryogenic freezing of the copper in the coil reduces (R) to about .003 Ω (*935 times less than normal*) intensifying the magnifying factor: Q_m = L/R = 2.06/.003 = 686.7 Q_m, almost a 700-fold increase. The cryogenic oscillator is incredibly efficient. It will resonate for several seconds with a brief power introduction. Low temperature does not cancel the induction and inductive reactance in the coil. Tesla's cryogenic oscillators need precision tuning between the transmitter and receivers.

Tesla's decision to apply cryogenics to his oscillator was expensive. It wasn't until Carl von Linde patented a procedure to liquify gas in 1901 that Tesla could reasonably apply the method to the Wardenclyff oscillator. Linde was a friend of Tesla's, and he coached Tesla in the art of multi-stage cooling. Tesla, in 1901, applied for and obtained patent 685012, applying cryogenics to his resonating transmitter, and in 1900, patent 655838 for insulating cryogenic conductors. It begins to come clear where Tesla's reasoning was going with these patents. Adding

---

[102] https://teslauniverse.com/nikola-tesla/patents/us-patent-655838-method-insulating-electric-conductors
[103] Colorado Springs Notes: page 202 – modified secondary

cryogenics to his magnifying transmitter removed two major barriers: the electrical resistance and parasitic capacitance. Tesla began to realize the possibilities of his dream, a world-wide communication and power network.

Figure 6.3 Wardenclyff Tower with the power and communication house, the antipode is in the South Pacific

Tesla was not one to 'give away the store' in his public patent ideas. His patents were intentionally vague. For instance, it was the idea that he was patenting in his bucket full of liquid air. No one could keep enough liquid air in a bucket to freeze a dozen fish, much less a heated coil fifty feet in diameter. Tesla built his resonator primary and much of the secondary of hollow copper tubes (patent 655838). He used his rotary pump (patent 1,061,142) to whisk liquid hydrogen through the plumbing creating a supercooled/supercharged system. Without the friction of resistance or parasitic capacitance losses, Tesla's system could transfer great power. Thus, due to the lack of friction in the oscillator, the power in the LEWs built quickly by standing wave resonance from antipode wave reflection. Standing wave voltage placed extreme strain on Tesla's oscillator coil and cupolas insulation. The high voltage would also back-feed into the primary input winding and burn out the primary power supply. Power regulation was an important part of the oscillator.

A resonant transceiver with almost no resistance in the primary or quarter wave coil can receive power from the environment. LEWs created by lightning strikes anywhere on earth within the range of the resonant frequency of the quarter-wave coil would feed energy into it. These LEWs over time cause a resonant power buildup in the transmitter, intensifying into a dangerous power breakout, possibly damaging the transmitter and endangering life. Resonance begins quickly with a standing wave established between the lightning bolt and the quarter wave transceiver (*transmitter or receiver*). The oscillator will go out of control within a few seconds after cryogenic freezing begins. A switchable load or cut-off switch in series with the ground wire can be installed to prevent overloads and to provide safety.

Lightning strike LEWs resonate with a transmitter and receiver tuned to the same frequency. In a transmitter only scheme, lightning induces power into the transmitter. The transmitter then echoes the power back toward the lightning bolt. Resonance keeps the lightning bolt active for a longer time. Interrupted flashes in the lightning bolt, displays the evidence of resonance power discharge from the standing wave.

Power in the LEW resonates with the charged cloud through the lightning bolt (*just as an antipode does*) for three or more seconds until friction weakens it. The lightning bolt will eventually cease and the transmitter will begin to resonate normally with its antipode using the remaining power. Continuous lightning strikes of the same resonant frequency anywhere on earth will feed the oscillator. The saving grace to a cryogenic oscillator is standing wave interference between geographically diverse lightning strikes. The LEWs have different timing hence cancelling-out most power surges.

Arrangements with transmitters and receivers have different challenges with lightning. The terminal spheres of the resonant pair gain similar potential voltages and are forced out of phase. The transceiver with the highest potential cancels the lesser one leading to power loss after the bolt disappears. Lightning strikes are strange and inconsistent, they even resonate with other strikes across the globe obstructing the opportunity for power harvesting.

In 1901 it was inconceivable to most everyone that Tesla applied cryogenic freezing to his large primary and secondary coils, and he did it on the hush. The Wardenclyff secondary coil was 51 feet in diameter with 17 windings. The tuned frequency of the frozen oscillator was a sharp, 90,140 Hz quarter-wave frequency not including a copula or extra wire length to earth ground. Tesla's oscillator performed between 65kHz - 90 kHz and inefficiently without the cryogenic super-conductivity. The use of cryogenics was a brilliant move on Tesla's behalf; it brought his world power and communication plan from 'maybe' into one of certainty.

Receivers

Tapping the power of the LEW demands a thorough understanding of reactive power and resonant arrangements. There are various transceiver designs. The application defines the style receiver required. First, in figure 6.4 is shown the basic, reactive receiver used in several of Tesla's patents.

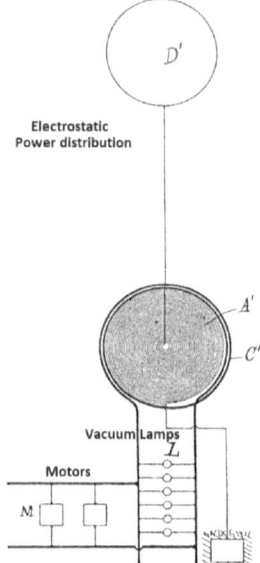

Figure 6.4 The basic Tesla resonant receiver

The detail described in his patents are skimpy but accurate. Tesla did not discuss how to convert high frequencies to usable AC power, it wasn't part of the patent, only the method of receiving the power. The receiver shown includes motors (M) and lights (L). Tesla's vacuum tube lights were high lumen, low-power, 'daylight bulbs. They were lit by high frequency electrostatic energy which would not shock or maim. The motors were also electrostatic; they lacked the power of his magnetic field motors, but they could do significant work. Tesla had the knowhow and technology to convert the [104]power at the receiver to normal sixty hertz voltage.

The receiver's secondary is the coil labeled (C') in 6.4. The primary (A'), is a quarter wave pancake coil. Establishing a standing wave LEW as shown in figure 6.5 calls for connecting a robust ground connection in the highest voltage swell of the LEW for best reception.

Receiver planted in a LEW crest

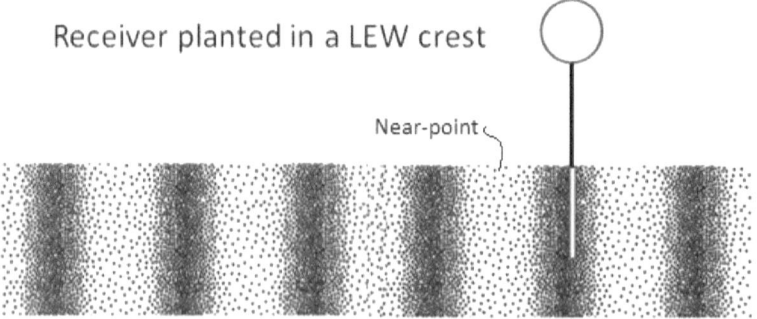

Near-point

---

[104] https://teslauniverse.com/nikola-tesla/patents/us-patent-390820-regulator-alternate-current-motors

Figure 6.5 A Tesla RPT receiver in a LEW compression point

Figure 6.6 details radial LEW movement away from and toward the ground plate of a transceiver. Near-points (*figure 6.5*) have zero voltage and are between complete cycles (*360°*). Half-points are also zero volts and are at half wave (*180°*) *points*. Both near-points and half-points are 1/8 wavelength (*45 degrees*) from LEW maximum voltage points. To calculate the distance of a near-point from the origin, divide the distance from the origin by LEW wavelength.

There are two nodes and two antinodes in each wavelength (*one near-point and one half-points*). A receiver placed on the crest of an antinode (*fig 6.5*) gives the best power. A 250,000 Hz LEW has a wavelength of 3,928 feet (*186000 mi/s/250000 Hz = 0.744 mi.*). Near-points are at 0.744-mile intervals from the origin. The transmitter is always a ground (*'0' Volt*) node which is the origin point.

Top view of a spreading LEW

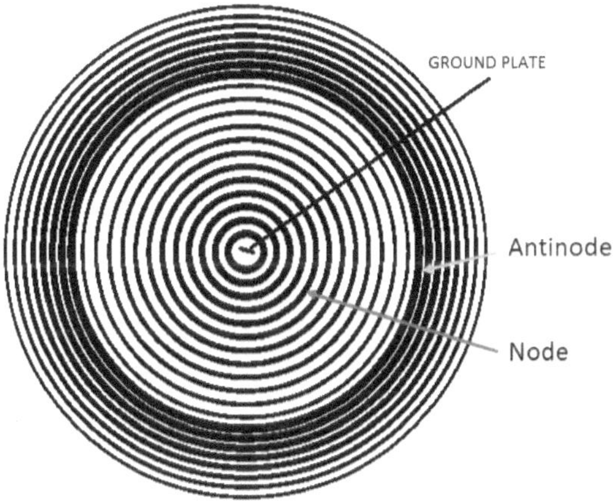

GROUND PLATE

Antinode

Node

Figure 6.6 Circular nodal points from the ground plate

Let's say we wish to place a receiver about 372 miles from the transmitter. Divide 372 miles by the wavelength of .744 miles, or 500 LEWs (λ). I will use full waves to avoid confusion. Our receiver will be placed near the 500th and/or the 501st full cycle LEW in an antinode near the 372-mile point. The 372-mile point is a near-point with zero volts. The receiver should be placed at 1/8 wave length distance or 491 feet (*.093 miles*) from the near-point. This is the closest antinode location.

The best method for placing a receiver is to use an aerial signal detector. Move about until reaching a maximum voltage reading of the LEW. At long distances, the spherical measurement of the earth (*see figure 6.7*) can cause miscalculations. The LEW can take a direct path to the receiver under the earth's surface at times. The 'as the crow flies' nature of the LEW (*figure 6.7*) can cause shorter distances than surface measurements (*remember Tesla's 2*π/2 it accounts for subsurface conduction*).

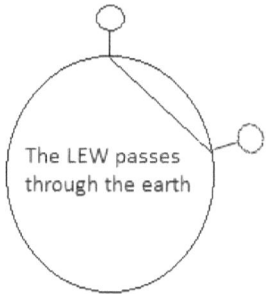

The LEW passes through the earth

Figure 6.7 The LEW short cut through the earth

The simplest and the oldest receiver style is in figure 6.8. The ground plates are placed in the LEW at node and antinode points. The receiver collects power from resonant or non-resonant LEWs. Ground plates placed at a node and antinode receive the LEW at the calculated wave length. Non-resonant LEWS create a sine wave signal. The resonant LEW receiver produces a pulsing, negative DC. Direct charging of batteries is possible with a resonant LEW receiver.

Earthed between a node and an antinode

Output to inverter

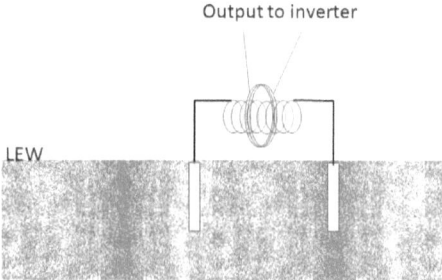

LEW

Figure 6.8 Two-Point Non-resonant, Standard Consumer Power System

Earthed at an antinode with aerial receiver plate

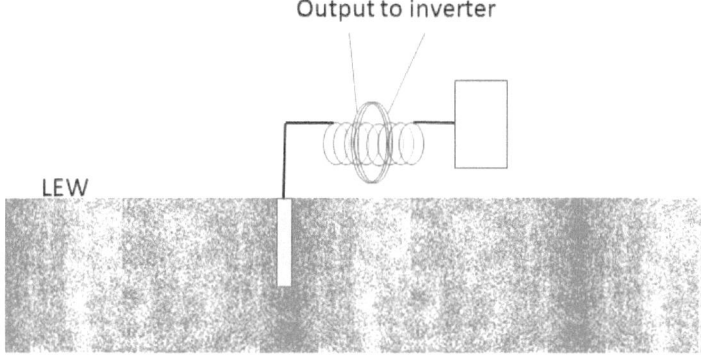

Figure 6.9 High power delivery for Home and business

Figure 6.9 is a resonant, reactive receiver with both power and communication abilities. Power delivery demand defines the scale. Power demand determines wire gauge, frequency, and terminal capacitor size. Parallel-wired terminal capacitors (*vacuum type*) should be used for high power demand. This receiver is like a standard resonant receiver. Locating the terminal over a node increases power production. The electrostatic induction increases current through the receiver coil. The receiver requires at least one-half wavelength between the ground plate and the capacitive terminal. The lead wire from the ground and to the sphere is part of the quarter wave coil length.

Earthed at an antinode with a timed switching
device to extract power from the LEW

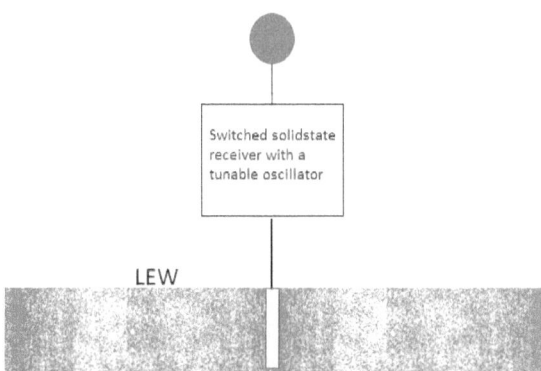

Figure 6.10 A simple switched single pulse receiver – requires power to begin operation

The last example in figure 6.10 is a switched receiver. Synchronization between the switching interval and the LEW frequency controls the efficiency. Power is controlled by terminal capacitor size, frequency, current capacity of the switch, and wire size. It works in a push/pull fashion for maximum power. This type receiver provides power for homes or businesses. Broadband Internet, VOIP, and supervisory communication are part of the switched arrangement. Various variations of this receiver are useful in rural and remote settings.

I cannot stress the importance of a solid ground connection in a terrestrial receiver. Maximum power availability is at the antinode crest point. Knowledge of the frequency and origin must be known to find the crest for ground placement. Reception of power needs a steady frequency from the transmitter, but there are exceptions. Switchable grounds placed at various points can receive various frequencies. A 'smart' receiver samples potential from various connections and connect them to the switch correctly for steady power.

Correct and incorrect ground connection points on a LEW

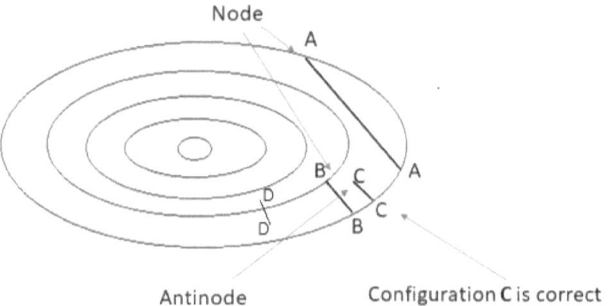

Figure 6.11 Valid and invalid LEW tap locations for dual plate systems

In figure 6.11 solid lines define the nodes (*positive potential*). Spaces between the lines (*antinodes*) are crests. They are negative in potential when measuring from the lines (*nodes*). Voltage varies according to closeness to the halfway point (*compression wave peak*) between node lines. Maximum power is between a node and antinode at points **C-C**. Lines **A-A**, **B-B**, **D-D** have the same potential and produce zero volts. The LEW provides power between any solid line and crest in the figure. The potential drops linearly between the antinode (*crest*) and the node.

Due to the nature of the LEW with its peaks (*antinodes*) and valleys (*nodes*), points exist where little or no power is available. A multiple frequency arrangement (*array*) is used to avoid signal loss because of low power. However, multiple frequency transmitters cannot exist within the same ground field because the frequencies will combine into one single frequency. The Fourier transform defines the single combined frequency of two or more co-grounded transmitters.

Arrays are a broadcasting arrangement of multiple frequencies. Each array transmitter has its own frequency and ground field. Arrays can provide three phase power LEWs. Each member of the array has different nodes and antinodes. Arrays improve EP wave density over that of a single wave LEW. The Array keeps the wave pattern tight and saves money in servicing time. The transmitters can be a few hundred feet apart.

The longitudinal wave pattern in figure 6.12 is an array with three frequencies. This is a Complex. The transmitters in the array have independent grounds to create this effect. Three transmitter arrays are arranged in a line or a triangle formation. Odd and even harmonic frequency spacing are not practical. A preferred technique is to increase the frequency by a few hundred hertz in each transmitter. [105]The formula y(n) = f[x(n) + c] where c is a small constant (*distance between transmitter ground fields*) other than zero to offset the array frequencies. Transmitter tuning must be precise. Wave symmetry interference can be prevented in other ways; see the [106]Stanford website in the footnote on this page for indepth information.

Sine wave power supplies are the preferred. Square wave power can decrease LEW voltage because of loss at upper harmonics. They also cause harmonic distortion in nearby transmitters. Design engineering is a must, but Field testing is necessary before plan completion. One more thing to consider is earth conduction anomalies. Poor conductance causes LEW distortion beyond the array.

Three phase, resonant array receivers should placed at a crest of a chosen LEW frequency from the transmitter array. Receivers should have individual ground fields, placed at a crest of the LEW. Using a common ground or placing array receivers within the same ground field will change the resonant frequency of the echoed LEW. Array receivers placed on the same ground will receive half power on its tuned frequency. The array will echo on a different frequency due to Fourier transform. This is an interesting and open doors to power relay plans converting three phases into one huge carrier LEW.

Multi-frequency transmitters in a field separated by a quarter wave of the lowest frequency

---

[105] https://ccrma.stanford.edu/~jos/pasp/Enhancing_Even_Harmonics.html
[106] https://www.slac.stanford.edu/econf/C9707077/papers/art42.pdf

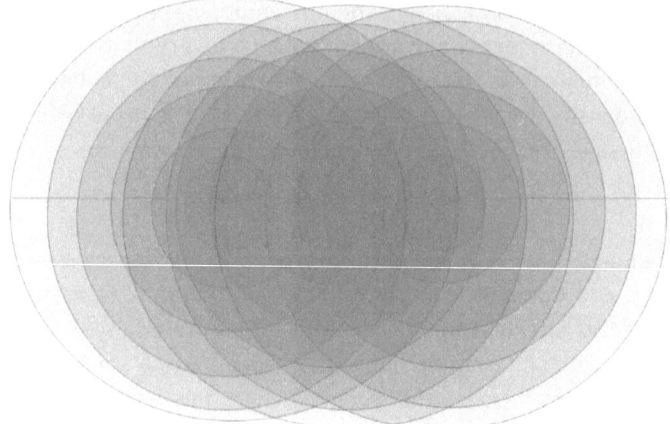

Figure 6.12 Complex Multi-standing wave arrangement with three transmitters

Figure 6.13 is a three-resonator array LEW pattern with a linear transmitter layout. There are no regions without LEW coverage in this array. The array assures that mobile receivers and aerial vehicles have consistent EP wave power. Mobile units may tune to all the frequencies simultaneously for augmented power. There are no dead spots in the range of coverage, there are no zero voltage points, but there are warmer and hotter spots in the LEW and EP field.

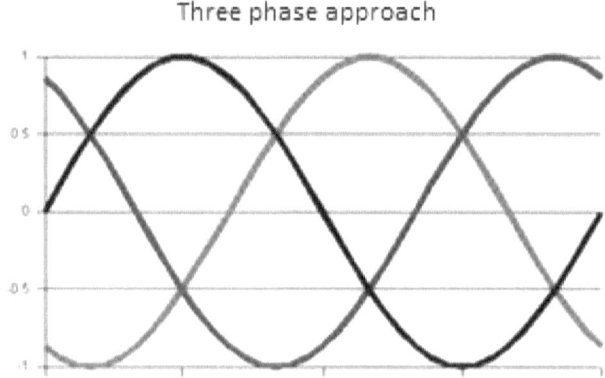

Figure 6.13 - Three phase signals sent from an array with three transmitters

The RPT transmitters can be encapsulated and submerged in salt and fresh water. Light speed communication, power transfer, and broadband internet are possible. We tested transceivers with two plates underwater. LEW power transfer, and resonance was stronger than earth-based systems. The full range of the underwater LEWs is still unknown. Underwater signal range is proportional to the voltage and frequency just as in the earth LEW transmission. We plan to do further underwater tests. The underwater LEW places no life in jeopardy and does not cause electrical shock. Signals pass equally well between earth and water with no loss at the boundary.

Imagine what long-range underwater communication and power transport to underwater drones would mean to science. An oceanographer could place a drone in the water, put on his 3D visor with earphones, and dive deep, traveling anywhere, studying or exploring. This technology would change oceanography like the Internet changed the world. A university in Boston already has a set of our custom-made resonant transceivers and preliminary testing has been successful.

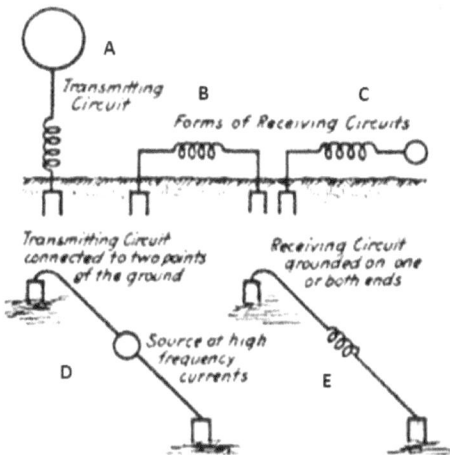

Figure 6.14 Various early forms of transmitters and receivers drawn by Tesla

LEWs travel below the earth's surface down to 21 miles deep. A specially designed RPT receiver can work underground. The receiver needs an electret (*a permanent magnet equivalent in electrostatics*) on one plate of a two-plate receiver (*Figure 6.15*). The receiver works in a cave or an underground void. Varying voltage on the non-electret plate creates a voltage difference across the receiver. Receiver orientation (*up, down, sideways, etc.*) doesn't matter.

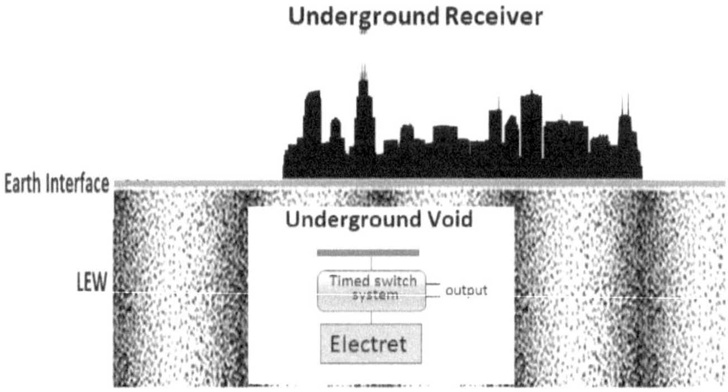

Underground Receiver

Figure 6.15 An underground receiver in a void

I will end this chapter with various forms of receivers and transmitters designed and patented by Tesla. Two of the designs in figure 6.14 had no patents. The forms 'A', 'B', and 'C' are Tesla's resonant transmitter and receiver. 'B' is a non-resonant receiver that extracts power from the LEW, and 'C' is a resonant receiver style.

In figure 6.14 above, 'D' and 'E' is an early non-resonant, damped LEW transmitter and receiver pair. The high frequency supply clamps directly to the sending plates in the ground. Power becomes lost between the ground plates as the frequency lowers. The power is supplied from a directly connected multi-pole generator, not a spark gap power supply.

Transport in example 'D' consumes significant power and has a short range. LEWs created by this design are full wave. Generator frequency defines the wave length. The example 'E' receiver is placed parallel to the transmitter. There is one LEW compression and one LEW rarefaction per cycle. The LEW waveform is interfering, concentric waves. Like waves created in a quiet pond by two stones thrown a few feet apart. The arrangement is inefficient and will kill or drive away underground animals.

Chapter 7 Wireless Resonant Systems Gotchas and Miscellaneous Information

The intent and focus of this book are to introduce new information. LEWs, EP waves, and power transfer through the natural medium are a new (*one-hundred-year-old*) science. The goal of this book is to provide a launch pad for continued research and to introduce common terminology. Reactive Power Transfer (*RPT*) taught here can apply to other sciences. I have held back very information little in these pages. I guarded future Patent information and that is all. LEW and EP waves could have full volumes and I plan to elaborate in the future on these subjects.

This final chapter is a lateral reference about electrostatics and EP fields. One can drift from the focus when writing about subjects with many facets. This chapter written to pick up some important scraps of information that fell on the floor in previous chapters. These scraps are ideas, reasoning, and/or logic.

Resonant signal collapse:

Resonant signal collapse (*RSC*) is a condition of signal loss between the transmitter and its resonant receiver/s. When the resonant signal is weak or disappears, the standing wave collapses. The EP field will fall apart as the LEW fades or fails. The condition that caused the RPT failure must be alleviated before reinitializing the standing wave.

QOS (quality of service) of a wireless power network should be better than local power service. Contractual agreements guaranteeing service are written to refund money after extended outages. Wireless power has no transport wires, poles, or transformers to burn out so a high QOS is important. Local RPT schemes such as a small 150-kW farm arrangement powering water pumps, electric fences, and lighting are not as critical, but are also subject to RSC. Power supply outages at the transmitter, circuit breakers kicked out, or loose wires at the receiver are simple repairs. More complicated events can happen.

When the Ionosphere is exposed to sunlight, the sun's photons cause a chemical release of electrons (*see foot note 20 in chapter. 5*) in the upper atmosphere. Closer proximity of these electrons to the positively charged ions in the Ionosphere causes the lines of force to switch to them (*diurnal switching*). The connecting force lines to the earth release. Switching electrostatic of lines of force triggers a ten percent loss in EP field flux density in the earth's surface electrons. Marginally powered systems may fail from an event like this.

An RSC event in Low voltage transceivers (*below 50 kV*) are particularly susceptible to an RSC during diurnal conditions. Adjusting to a higher output voltage averts the condition. As you know now, the Tesla RPT system relies on electrons as a medium. A dense medium conveys more longitudinal power. Increasing voltage on the transmitter in daylight hours strengthens and maintains the EP wave power delivery. Bear in mind, the transport path is reactive, it does not consume power but returns the power every half cycle, making a power increase a fully refundable investment.

Various causes of (*RSC*) Resonant Signal Collapse:

1) Diurnal RSC due to electron migration back into the earth's subsurface
2) RSC caused by high ground resistance at the grounding plate or rod
3) Trouble in the capacitive terminal due to open wire, broken tube, or eroded protective covering
4) A flooded ground field in a three-frequency array causing the individual transmitter nodes to become a common node connection causing a standing wave frequency change
5) Earthquakes can create excess surface electrons and cause a LEW reflection at the densest points in the electron field reflecting the standing wave. The excess electrons from a tectonic compression area also attract lines of force from the Ionosphere weakening the EP field.
6) Lightning strikes to earth near the transmitter or receiver can interrupt the LEW momentarily
7) Other miscellaneous reasons

While I was compiling information for this book, I read several references about Tesla's flying craft, not with propellers, but the one that he mentioned in a [107]b ook by Cheny. Tesla said, "You should not be at all surprised, if some day you see me fly from New York to Colorado Springs in a contrivance which will resemble a gas stove and weigh as much. ... and could, if necessary, enter and depart through a window." (7-7-1912) [it will be a small box, not a huge "cigar"]. He also spoke of gravity shielding, but there are no actual patents or publications, there is simply nothing factual in books or the WWW describing the principle, and I looked thoroughly.

I dug through various references for information focusing on Gauss's first law of electrostatic forces. Gauss states the force can be calculated between charges. The formula is $F = q_1 q_2 / 4\pi\mathcal{E}_0 R^2$: where q is the charge, $4\pi R^2$ is the area of a sphere (*charges are spherical*), and $\mathcal{E}_0$ (*epsilon naught*) is the permittivity of free space. The R^2 in our equation tells us that the negative or positive forces decrease according to the square of the distance between them.

Charges exist on the surface of a sphere. There are no charges inside the sphere. If a large sphere, of 9 meters in diameter weighing two tons, had a negative charge placed on its surface. What charge density would be required to produce a force to lift it? This is a two-part question. With a negative charge, the sphere would be pushed away from the earth at a rate of speed according to the coulomb

---

[107] Pg. 198 Tesla, Man Out of Time by Margaret Cheney.

force on the sphere. Simultaneously the Ionosphere is acting to attract the sphere. There exist a push and pull force on the sphere. The electrostatic forces would have to exceed the weight of the craft. Lift comes from charge, not airfoils. Increasing the electric field on the surface of the craft should, in theory, initiate lift-off. With a ten-million-volt negative charge, the sphere would glow like a ball of lightning with an electrostatic aurora. Heavier craft requires more charge. Maybe high electrostatic charges negate, shield, or disturb the gravity field? I plan to study this closer.

Tesla would have applied a negative pulsing generator inside of the sphere. A Tesla coil secondary with one end connected to the sphere and the other one left unconnected was a motif Tesla used in electrostatic experiments. How much negative charge potential would be required to lift this sphere? Let's assume the sphere weighs 830 Kg (*about 2500 pounds*). You do the math. Isn't it curious that in the evening there is an increase of UFO sightings? These may be Tesla driven, electrostatic craft.

One problem, any electrostatic craft relying on earth repulsive force for more than 50% lift would crash to the ground with a strong, rarefaction LEW caused by a lightning strike. For safety sake, reliance on the Ionosphere for pull is more important than relying on the fickle earth electrons for a push. Tesla needed a prime mover to fulfill his plans. The 'Prime Mover' was the light and simple [108]Tesla turbine and it produced a profusion of power.

In recent years there have been various trials of wireless power transfer systems using magnetic resonance coupling, plain-vanilla magnetic coupling, radio waves, and microwaves. There was an interesting study in the nineteen-nineties with high frequency (SWEPT) [109]Single Wire Electrical Power Transmission. It was 'all the rage' for a short while. The system was a copy of Tesla's patent 593,138 from 1897. This device was demonstrated on a 20:1 scale at the 2019 ESTC conference in Idaho by Eric Dollard. The two transceivers were beautifully crafted. It was an impressive display of the power in Tesla's design.

---

[108] http://www.tfcbooks.com/articles/tdt7.htm
[109] https://file.scirp.org/pdf/ENG20121100002_50196856.pdf

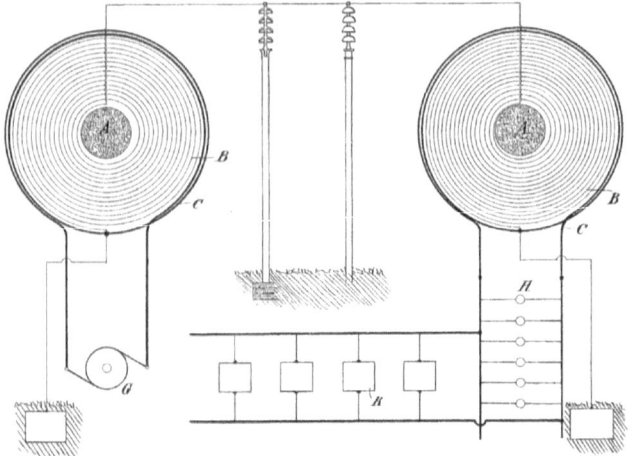

Figure 7.1 Tesla patent 593,138

Tesla used high frequency power from a multi-pole generator (*10 kW at 10kHz*). The power transferred by transformer action into a quarter-wave-length pancake coil secondary. The original arrangement used earth as a return wire. This is the patent idea used at the Niagara Falls power station. The high cost of power line construction started Tesla's mind to thinking of ways to reduce cost. It was similar lines of progressive thinking that eventually steered him to drop the wire all together and use earth ground as a conductor. Tesla knew his patent did not need the ground wire on the high frequency power transformer arrangement but it remained on the design.

Tesla knew the medium for power transport through the earth was the interstitial electrons brought to the surface by the positively charged atmosphere. His tests proved electrons have a high bulk modulus (*"electrons are incompressible"*, *in his words*) making them the perfect medium for electron soundwaves. The formula for [110]Bulk modulus is $B = \Delta P/(\Delta V/V)$, where P is pressure (*voltage*) and V is volume, a higher bulk modulus is less compressible. Electron have no elasticity intrinsically, therefore $\Delta V = 0$. **B = 0** is not the case because the surrounding charge of the electron *is* compressible, the reciprocal of bulk modulus. The ΔV figure is minute making an enormous B value.

The electron's surrounding charge cause them to repel one another but they are compressible by high-voltage. The compressible **E** field around an electron reduces the bulk modulus of the electron. A higher ΔV (*volume*) happens by the 'E' charge lowering the bulk modulus 'B'. A lower 'B' figure produces 'sponginess' in the medium. LEWs are longitudinal waves and demand a high 'B' value for efficient transport.

---

[110] http://hyperphysics.phy-astr.gsu.edu/hbase/permot3.html

For example, a five-volt arbitrary waveform generator can propel a 500 kHz LEW three hundred feet with an acceptable ground. LEW range is frequency dependent owing to high (X_L) resistance.

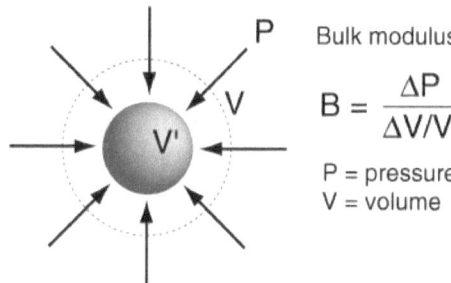

Bulk modulus:

$$B = \frac{\Delta P}{\Delta V/V}$$

P = pressure
V = volume

Figure 7.2 The bulk modulus of an electron

Similarities between earth and a wire conductor

In telephony, time division reflectometry (*TDR*) tests define wire or cable length. An electrical pulse inserted at one end travels down the wire to the terminal (*end of the wire*) and echoes. Several things define the time delay of the reflected pulse: the metal type, wire gauge, the insulation, and the temperature. Increased temperature in a conductor excites atoms. Hotter atoms and their electrons do not have a high a bulk modulus because of heat energy. The heat effect slows and weakens a TDR signal.

Conversely, the colder the conductor the closer atoms and electrons are together permitting a faster speed for the TDR compression signal to travel. The TDR voltage pulse is negative or positive. The pulse causes a compression or rarefaction wave to move in the wire. The electrons swing a minuscule distance depending on voltage, and return to their original position after the signal passes. Electrons 'bump' bordering ones in a chain reaction until the signal reaches the end of the wire to a dead-end and returns to the TDR meter.

The LEW is like a TDR wave in the earth conductor. Reiterating, Tesla knew about the LEW, though he called it a "type of sound wave". He described how it moved in the medium, and he calculated its speed. Tesla's words, paraphrased, *"Electrons are an incompressible fluid, that when disturbed, can provide a medium for waves that travel outward from the ground plate in concentric rings."* A Tesla hallmark was his ability to "think out of the box". I did not immediately grasp the implications of his statement about the LEW being a soundwave (*though he did not call it a Longitudinal Electron Wave*).

I worked on this a Tesla oscillator for two years, and it wasn't until the third year when I was researching underwater light speed broadband that I began to understand the LEW. I presented a lecture in Boston at Northeastern University in

May of 2016 on underwater reactive communication. Later that year, as I mentioned earlier in this book, I sent them a working prototype an RPT oscillator. They had successful experiments in their test aquarium, and ocean testing is on the agenda, but funding is a problem (*Deja vu*). In time, someone will devote time and money toward developing it.

A simple tuning for RPT transmitters and receivers:

During our tests we always found it difficult to achieve exact tuning between the transmitter and receiver. I found a quick and easy precision method to do it.

The items needed:
1) an isolation transformer
2) an oscilloscope
3) an arbitrary waveform generator
4) some wire (*to connect to the ground rod*)
5) two ground rods, or a couple of buried copper plates
6) two electrical adapters to convert from three (*with GND*) to two prongs (*without GND*)
7) a quarter wave receiver or transmitter to be the DUT (*device under test*). Figure 7.3 below shows the test setup

1. Drive the ground rod, or bury the ground plate for the waveform generator – it can be a plate in tidal water, or about four feet in a hole in damp earth, a rod should be long enough to reach moisture in the earth, usually six feet or more.

2. Connect a wire to the ground rod, and bring it into the work bench area where wave form generator can be connected to it

3. Connect the isolation transformer to a power source at the work bench

4. Connect a three-prong adapter to the output jack of the isolation transformer (*this is necessary because the ground from the power company is cut through to the isolation side of the transformer: the power company ground cannot be connected to the waveform generator or the test will fail*)

5. Connect the random waveform generator to the three-prong adapter which is plugged into the output of the isolation transformer with the ground lead disconnected

6. Take the positive (usually red) lead of the generator and connect it to the ground wire going to the ground rod or plate, this will be the transmitter's signal input to earth; leave the ground clip disconnected

7. Drive another ground rod, or place a ground plate, at least 100 feet (or 1/8 wavelength) away from the wave form generator's ground rod (the transmitter

and receiver cannot be in the same earth grounding plane or the signal will be shorted out)

8. Connect the receiver or transmitter quarter wave coil to its own ground rod or plate

9. Connect the oscilloscope across to the receiver/transmitter unit secondary winding, fig. 7A

10. Estimate the frequency of the receiver/transmitter device under test: multiply the length of the quarter wave coil by 4 and add the length of the wire connecting it to ground. Then divide that total length into 'c' (*186,000*). For example: If the quarter wave coil is 900 feet long, then the wave length is .68 mi (900*4 = 3600 ft. or (3600/5280=.68 mi.). Now, assuming a connecting wire length of 75 ft. then 75/5280 = .014 mi., total the wire length, .014 + .68 = .694 mi., finally calculate the frequency from the wavelength - (186000 mi/s) / (.694 mi) = 268,011 Hz.

11. Place a square wave or a sine wave signal from the generator in the range calculated in step j (*a square wave is more powerful than a sine wave*) into the ground wire connection

12. Turn on the waveform generator at maximum output and begin to sweep the frequency of the quarter wave receiving device downward or upward (*the starting range of the frequency determined by the length of the quarter wave coil calculated in step j*).

13. Watch for a signal on the output of the receiver station. You will either need help in this process, or use two cell phones with Skype or similar software to watch the distant receiver output while tuning the generator.

Various harmonics can be received, both high and low ones and they will be visible as the frequency sweeps up or down.

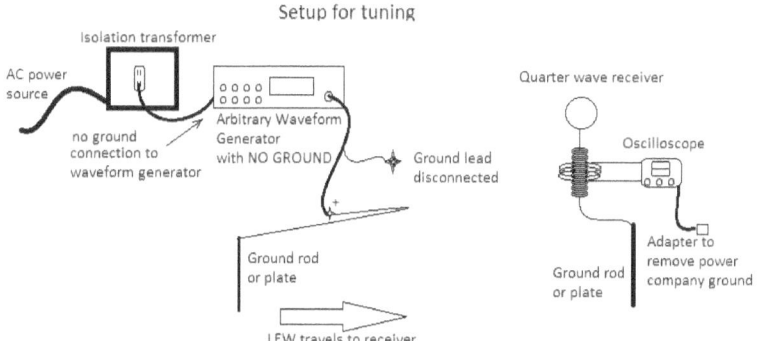

Figure 7.3 A Simple tuning arrangement for a Tesla RPT receiver

Waveform generator ground isolation is necessary. The waveform generator case and signal ground connections have common connections inside the casing. To do these tests the common ground circuit must be broken through the isolation transformer. This protects the signal from a short-circuit when the (+) lead connects to the ground rod/plate. This step assures the signal from the variable waveform generator will enter the earth ground and make a LEW.

The low voltage signal of the waveform generator (*5-10 VAC*) will reach up to five hundred feet through the earth. At the receiving end, the receiver's secondary coil should not be grounded. The oscilloscope may be connected to a regular power outlet. Power company ground isolation is preserved by the 'dry' secondary coil. Any ground condition on the secondary of the receiver will prevent signal reception.

Sweep through various frequencies with the signal generator until a signal is detected at the receiver. Secondary receiver voltage and the frequency is adjustable with the waveform generator at the transmitter. Various harmonic frequencies and voltage can be tested and experimented with. There is no resonant LEW power gain. Standing wave power from constructive interference is seen between the signal generator and the receiver. The standing wave has a 180° compression and 180° rarefaction cycle resembling a sound wave because no resonance occurs between the arbitrary waveform generator and the receiver. The received signal's strength reading is ambiguous unless properly calibrated as shown in figure 7.3-B.

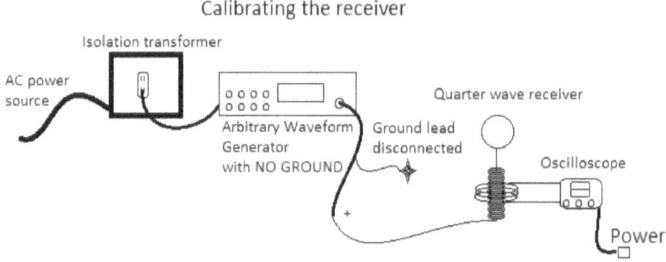

Figure 7.3B Calibrating the receiver

If calibration is desired, follow these steps.

1. Place the receiving device *on the bench* with the waveform generator, and connect the waveform generator's positive lead to the 'ground wire' of the device under test and leave the ground lead unconnected.

2. Place the oscilloscope across the secondary coil of the receiving device. Adjust the waveform generator's power to maximum level, and read the secondary voltage while sweeping the frequencies (*This system does not require a closed loop to operate, remember it is reactive. The electrons from the generator go through the quarter wave coil, charge the sphere, and return to the generator via reflection.*)

3. Tune the waveform generator frequency up and down until the resonant frequency is determined by signal amplitude (*this will be the frequency where the maximum voltage is on the secondary by the oscilloscope*). This is the **bench resonance** of the device, and it will vary slightly when transferred to the ground rod.

4. The 'bench mark' voltage level made at resonance is the **calibrated maximum output** with the waveform generator's maximum setting. With this benchmark test complete, you can now compare to the signal level during the generator tests through the earth to a known value. The ratio of the two signal strengths (*bench mark voltage vs actual working system readings*) can determine the loss or gain of the receiving device through the earth.

5. The transmitter and receiver cannot be energized in nearness because the 'E' and magnetic field coupling skews test results. Transmitter secondary voltage cannot be calculated by winding count, it varies by frequency. Many of the secondary windings do not have mutual induction with the primary.

Interestingly, there are ions at the earth's surface caused by cosmic bombardment that steal electrons to complete their valences. These ion captures can prevent electrons from contributing to LEW transport. *Local* barriers to LEW transport are ionic compounds, acid rain, hot springs, mineral crystals, and dense regions of frequency-resonant Rare Earth Elements (*REE*). Interference of LEW transport by these mechanisms result in creation of electric and magnetic fields. Whether good or bad for the LEW transport, I cannot tell. It may provide a detection method for pinpointing certain crystals or metals deep in the earth.

Tesla's secret power source...very possible

This is my own speculation about one of Tesla's biggest secrets never formally disclosed by him was free power. I believe he planned to use energy from the earth's lightning LEWs to power his telecommunication and power transfer towers at Wardenclyff. The original plan called for three towers; one of which was to be six hundred feet tall. That is sixty stories tall; a tall tower for a copula of .661850 Farad capacitance to undulate at a common earth lightning LEW frequency.

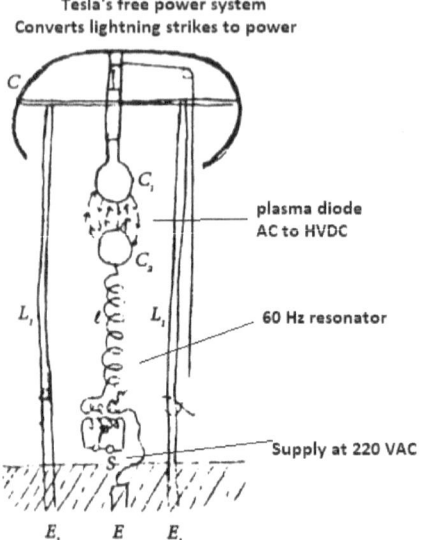

Figure 7.4 Tesla's power generator

I mentioned in an earlier chapter the lightning bolt length from cloud to ground defines the quarter wave frequency. The average bolt is five miles long. This calculates to a 20-mile wave length (*4x5mi = 20mi*) with a frequency of 9,300 Hz. Tesla did not use a quarter wave coil but metal stanchions connected to a very large copula to resonate at lightning LEW frequency.

The four stanchions were to be 600 feet tall (*figure 7.4*) with a three-square feet size minimum at the bottom. Tesla used the figure of four inches in diameter per support, at six hundred feet tall the tower would be rickety. Stanchions at 600 feet tall and three-feet-square have 1.77e+5 nH (*Nano Henrys or 10^{-9}*) or .00000000177 H. Placed in parallel, each pair (*they were shortened and spiraled in the final design*) would have .885 nH for a total of 0.4425 nH for four. The terminal would require a 6.6185e+5 microfarad capacitance to resonate at 9300 Hz. If Tesla used his vacuum tube capacitors in an array instead of a metal cupola achieving 9300 Hz. resonance would be no problem.

In figure 7.4 the arrangement in the center has cold cathode spheres linking them. The cupola coupling capacitance is adjusted with the pulley on the top sphere. The two spheres serve as a large spark gap creating a high voltage plasma that powered a lower Tesla oscillator. In his design, Tesla drew the secondary of the lower oscillator with a large 'S' for Supply instead of the usual 'G' for generator. Tesla directed the power from the NM to the powerhouse and used it to drive the other appliances and other oscillators. Tesla had large coal turbines installed for backup power, and later used them for the main power source for the communication tower. Tesla's dream three tower system was one power supply tower, one power transfer tower, and one communication tower. J. P. Morgan plugged Tesla's cashflow and he built only the communication tower with a compromised design.

The extra electrical capacity produced by the power tower of the original plan would have been wasted. Tesla decided he would give the power away to distant users. When he told Morgan about this plan, Morgan gave a flat NO and cut off all his funding. Morgan said he would have no part of giving away electricity.

Tesla had only $150,000 from J. P. Morgan for a first installment to build his world power plan; he used it up, spent his personal money, and became deeply in debt. He made enemies in both Colorado Springs and in New York because he borrowed and spent more money than he had. He was in a terrible financial position, and he embarrassingly had to walk away, being responsible for many lost jobs, and unpaid bills.

Lightning LEWs

LEWs of many frequencies constantly transverse the earth's surface. Enough power is available in these LEWs to produce electricity for a small community. While lightning is happening somewhere every moment of the day on the earth, the many frequencies involved in various lightning signals make it difficult to capture enough for consumer use. A cryogenic, multi-frequency receiver is probably the answer to the problem. Tesla had several [111]patents in cryogenic conductor methods. The main reason for the patent was to remove friction and improve power transfer capability.

Tesla's first lightning receiver used a spark gap that activated in the presence of the strong electrostatic (EP) wave of the LEW. When the spark gap fired, a capacitor and a '[112]sensitive device' provided the crackling sound of the standing wave in the earpiece. Later he used the large Colorado Springs transmitter's primary as a[113]secondary winding with a Wheatstone bridge arrangement and a telephone set across it. The receiver was tuned to 93,467 Hz. It was during a silent time between lightning strikes that he heard voices on his receiver. The voice transmission came over an EP wave, not a transverse, radio wave. Tesla built his wireless power rig in such a way as to filter radio waves to improve efficiency. He termed the people he heard talking as Martians and he did not understand the language.

---

[111] https://teslauniverse.com/nikola-tesla/patents/british-patent-13563-improvements-and-relating-transmission-electrical-energy

[112] The sensitive device was used as a weak signal detector in early radio sets (*made of nickel granules in a tumbler*)

[113] The Colorado Springs secondary coil had a fifteen-meter diameter (49.2 feet) with seventeen turns (*2,628 feet long*). The wavelength was: λ = 4* 2628 = 10,512 feet or a wavelength of λ = 1.99 miles. The frequency was 186000/1.99 = **93,467** Hz placing it right in the range of average lightning bolts. See page 213 of CS notes.

Cryogenics

Supercooled transceivers have low frictional resistance (R). This dramatically raises the [114]Q factor and narrows the resonant bandwidth. Eliminating dampening friction ([115]L/R) increases the power burst into the ground plate, removes parasitic capacitance, and increases the LEW magnitude. To explain the advantage of low resistance in the transmitting coil just look at the figures.

Assuming the inductance of the quarter wave coil is 750 millihenries and the resistance is five ($R = 5$) ohms then Q = 0.750/5 = 0.15. Cryogenically cooled copper has almost a zero-ohms resistance. For example, assume that in a supercooled coil that R is 0.0003 ohms: recalculating the Q with the cryogenic value: Q = L/R: Q = 0.750/.0003 = 2500, a 16,666-fold increase using cryogenics. An added benefit, the transceiver's bandwidth sharpens to a near exact resonant frequency strengthening the LEW.

More about lightning LEWs:

The LEW from a [116]lightning strike can begin life as a rarefaction or compression wave. Lightning strikes from the sky to the earth produce LEW compression strokes. Usually lightning bolt begins at the ground and releases toward the sky creating rarefaction LEWs. The lightning bolt's length defines the resonant frequency. The bolt is the quarter wave antenna. In this natural arrangement, the cloud above serves as the spherical capacitor, the lightning bolt is the conductive quarter wave conductor, and the ground plate is the point of touchdown. [117]The average lightning bolt is five miles long. According to calculation, the LEW frequency produced by a 5-mile-long strike is: 186000/(4*5) = 9,300 Hz, and the wave length is 20 miles (*4*5 mi.*). The period of the bolt from earth to cloud is 1s/9300 Hz = $107.53*10^{-6}$s, or 107.5 microseconds.

We can harvest lightning. A multi-tuned *cryogenic* receiver could do the job. If free lightning power were made available, everyone will want to use it, but there is no free lunch. Like a river through the desert, this natural power source would be overdrawn quickly. Power consumption from many users would cause RSC (*Resonant Signal Collapse*) resulting in no power for anyone. The resource would require international regulation, and management could be done using a petroleum business model. Power is pumped-up, cleaned-up, encrypted to prevent theft, and sold at a reasonable price as a subscription service with no middlemen.

---

[114] https://en.wikipedia.org/wiki/Q_factor
[115] https://www.allaboutcircuits.com/textbook/direct-current/chpt-16/why-l-r-and-not-lr/
[116] http://www.weathergamut.com/2015/08/06/the-positive-and-negative-sides-of-cloud-to-ground-lightning/
[117] http://www.ira.usf.edu/cam/exhibitions/1998_12_McCollum/supplemental_didactics/25.Uman10.pdf

Tesla discovered another source of [118]reactive power he did not describe except by the word 'Cosmos'. A modulating cosmic field must have a reactive receiver to collect it. The question I have for Mr. Tesla, "Why didn't you tell us more?" Tesla's radiant energy [119]patent was the nearest thing to a cosmic receiver apparatus he ever mentioned. Tesla's radiant (*cosmic*) receiver didn't have enough energy density to provide power for heavy demand, at least in its patented form. The cosmic receiver panels are similar in size to solar panels but without the silicon and much cheaper to build. Power is available twenty-four hours not just on sunny days.

Tesla believed there were powerful waves passing all over the earth from space which could produce a lot of electrical energy. When he returned from Colorado in 1900, he wrote the paper [120]"**The Problem of Increasing Human Energy**". Tesla wrote speaking of a user of the power, "But by the means I have developed he (*a person receiving the energy*) would be enabled to concentrate the larger portion of the entire energy transmitted to the planet in his instrument, and the chances of affecting the latter are thereby increased many millionfold." Again, Tesla left out details or hints. Transverse waves can pass through a vacuum but transfer little power. Longitudinal waves require a medium so according to Academia and current physics it is not possible for a longitudinal wave to pass through the vacuum of space unless the longitudinal wave it is a type of modulated, gravity, electric, charged particle, or magnetic field. Science tells us there is no 'etheric' medium and that space is a vacuum, but there is likely an undiscovered avenue of electrical transport we are yet to learn of.

Tesla's 'great' power source discovery was probably a longitudinal power wave with a frequency such that electrical properties of the earth are used to receive it. Free space has a resonant frequency defined by ε_0 and μ_0 (*epsilon naught and mu naught*). $\varepsilon_0 = 8.85 \times 10^{-12}$ Fd and $\mu_0 = 1.257 \times 10^{-6}$ H. These two constants also define the speed of light through a vacuum according to Maxwell's equations. By using these two figures, the resonant frequency ($F_R = 1/2\pi \sqrt{LC}$) of space is 47.718 MHz. The wave length is 20.5 feet with a five-foot quarter wave. A reactive receiver tuned at this frequency could possibly receive power limited only by the scale of the receiver. A Tesla reactive receiver is required to connect to the modulating power.

This world would be a different place today if Tesla's dreams of worldwide communication (*type of early Internet*) and power distribution were realized. The man created our modern society with his inventions and ideas. Thank you, Mr. Tesla, we owe you more than we can pay. We have a lot to learn.

---

[118] This word is my own addition: Tesla did not use the term reactive power.
[119] https://teslauniverse.com/nikola-tesla/patents/us-patent-685957-apparatus-utilization-radiant-energy
[120] http://www.tfcbooks.com/tesla/1900-06-00.htm :.

THE END

Glossary of terms

- **Antinode:** a maximum negative voltage or compression wave peak
- **Antipode:** a point directly on the opposite side of the earth from the transmitter
- **Bulk modulus:** The bulk elastic properties of a material determine how much it will compress under a given amount of external pressure. The ratio of the change in pressure to the fractional volume compression is called the bulk modulus of the material.
- **Complex LEW:** A two or more phase LEW arrangement (usually three phases for commercial power provisioning)
- **Concurrency:** When a Tesla transmitter and receiver are located within the same node
- **EP:** Electron plasma, defined at the electric and static field at right angles to a conductor or perpendicular to the NM, and can contain charged particles
- **LEW:** Longitudinal electron wave created in the NM by an RPT system
- **LEW Attenuation Table:** Defines the geographical locations on earth where electrons are rarefied
- **Longitudinal wave:** A wave that moves latterly forward and back like a spring coil and not transversely as radio waves. Longitudinal waves move via compression and rarefaction.
- **NM:** Natural medium – earth, water, and air
- **Node:** a point of zero volts potential in a LEW – it is also positive in respect to LEW maximum potential which is always negative in a resonant longitudinal wave
- **Node concurrency:** A condition where the transmitter and the receiver are in the same ground field.
- **Quarter wave:** one fourth of a full wave, a 100-foot wave would have a quarter wave of 25 feet
- **Reactive Power:** In electric power transmission and distribution, defined as volt-ampere reactive (var) is a unit by which reactive power is expressed in an AC electric power system. Reactive power exists in an AC circuit when the current and voltage are not in phase.
- **Resonance:** The building in power of a standing wave in a medium by constructive interference
- **Reflection:** the echoing of a LEW wireless signal from the antipode, or from a resonant receiver
- **RPT:** Reactive Power Transfer, the Tesla method of moving power through the earth or NM with negligible power loss in transport
- **Scalar wave:** A wave that travels independently of a time varying field and is not defined by the limits of permeability and permittivity (mu and Epsilon naught)

- **WPT:** Wireless Power Transfer, the moving of power from one point on the globe to another via AC currents creating LEWs. The reactive transport of oscillations from one resonating system to another without power loss.

INDEX

Bibliography

Nikola Tesla, Colorado Springs Notes 1899-1900, New York, NY

Anderson, Leyland, editor, Tesla On His Work With Alternating Currents , Breckenridge, CO

Anderson, Leyland, editor, Tesla Guided Weapons and Computer Technology , Breckenridge, CO

Nikola Tesla, Experiments with Alternating Currents of High Potential and High Frequency,

DAVID H Childress. The Fantastic Inventions of Nikola Tesla (p. 1). Kindle Edition, Stelle, IL

The Nikola Tesla Museum, Tesla's Manuscripts, Krunska, Belgrade

Nikola Tesla, My Inventions from Electrical Experimenter Magazine, dminoz.com

The Earth's Electrical Surface Potential, A summary of present understanding by Gaetan Chavalier, PhD

Nikola Tesla, The Nikola Treasury, New York, NY

Michael W. Simmons, Nikola Tesla, PROPHET OF THE MODERN TECHNICAL AGE, Kindle Books

William R. Lyne, OCCULT ETHER PHYSICS, Kindle Edition, Lamy, NM

Nikola Tesla, Phenomena of Currents of High Frequency, LEEAF.com

Nikola Tesla, The Problem With Increasing Human Energy, NY, NY

George Trinkaus, Tesla the True Wireless, Unknow Location

George Haller, The Tesla High Frequency Coil, Kindle Edition

W. Bernard Carlson, Tesla Inventor of The Electrical Age, Kindle Edition

Roy Fowler, Tesla's Gravity Engine, Amazon Books

Michael Faraday, Experimental Researches in Electricity Vol 1

Robert Haralick Ph.D. ESTC 2019 presentation on Longitudinal Wave solutions Haralick.org

Appendix A

EXCERPT from TESLA – found at
http://www.tfcbooks.com/tesla/nt_on_ac.htm#Section_4

Tesla's SCALAR WAVES:

The law which I discovered in Colorado is wonderful, and it shows that results undreamed of before and of incalculable moment will be obtained as soon as a plant, embodying these principles, is established on a large scale.

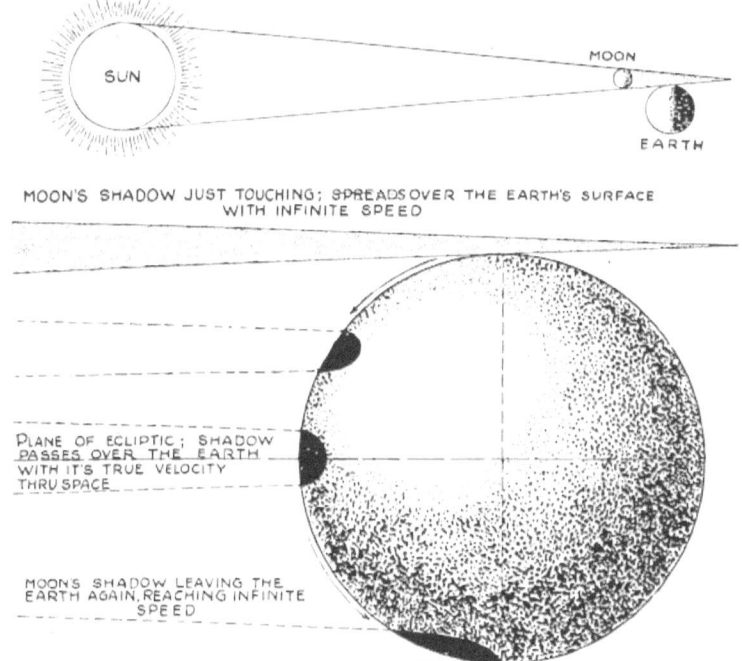

Figure 81.

Diagram illustrating the mode of propagation of the current from the transmitter over the earth's surface.

IN TESLA's OWN WORDS

To give you an idea, I have prepared a diagram [Fig. 81] illustrating an analogue which will clearly show how the current passes through the globe. You know that in a solar eclipse the moon comes between the sun and the earth, and that its shadow is projected upon the earth's surface. Evidently, in a given moment, the shadow will just touch at a mathematical point, the earth, assuming it to be a sphere.

Let us imagine that my transmitter is located at this point, and that the current generated by it now passes through the earth. It does not pass through the earth in the ordinary acceptance of the term, it only penetrates to a certain depth according to the frequency. Most of it goes on the surface, but with frequencies such as I employ, it will dive a few miles below. It can be mathematically shown that it is immaterial how it passes; the aggregate effect of these currents is as if the whole current passes from the transmitter, which I call the pole, to the opposite point, which I call the antipode.

Assume, then, that here is the transmitter, and imagine that this is the surface of the sea, and that now comes the shadow of the moon and touches, on a mathematical point, the calm ocean. You can readily see that as the surface of the water, owing to the enormous radius of the earth, is nearly a plane, that point where the shadow falls will immediately, on the slightest motion of the shadow downward, enlarge the circle at a terrific rate, and it can be shown mathematically that this rate is infinite. In other words, this half-circle on this side will fly over the globe as the shadow goes down; will first start at infinite velocity to enlarge, and then slower and slower and slower, and as the moon's shadow goes further and further and further, it will get slower and slower until, finally, when the three bodies are on the plane of the ecliptic, right in line one with the other in the same plane, then that shadow will pass over the globe with its true velocity in space. Exactly that same thing happens in the application of my system, and I will show this next.

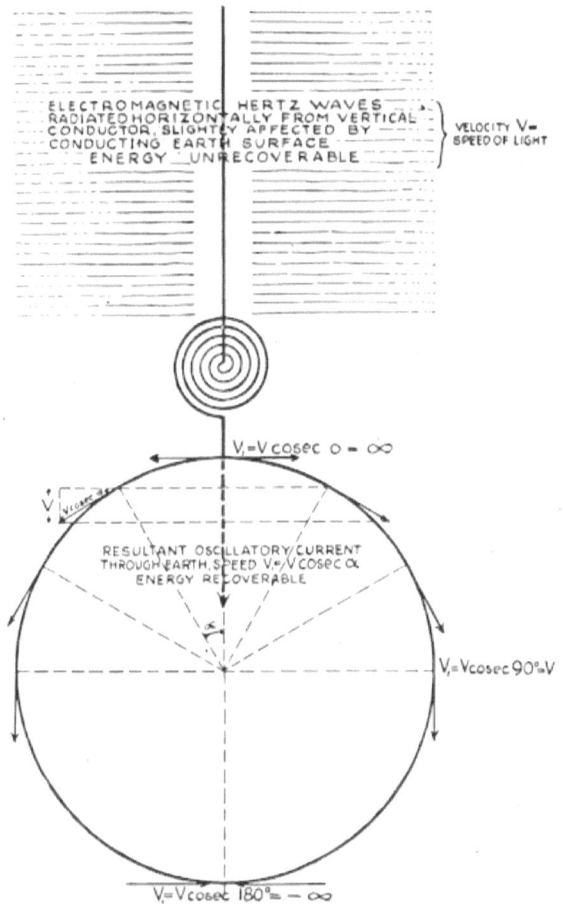

ELECTROMAGNETIC HERTZ WAVES
RADIATED HORIZONTALLY FROM VERTICAL
CONDUCTOR, SLIGHTLY AFFECTED BY
CONDUCTING EARTH SURFACE
ENERGY UNRECOVERABLE

} VELOCITY V=
SPEED OF LIGHT

$V_1 = V \cosec 0 = \infty$

RESULTANT OSCILLATORY CURRENT
THROUGH EARTH, SPEED V=V COSEC α
ENERGY RECOVERABLE

$V_1 = V \cosec 90° = V$

$V_1 = V \cosec 180° = -\infty$

Figure 82.
Diagram illustrating the law governing the passage of the current from the transmitter through the earth, first announced in U.S. Patent No. 787,412 of April 18, 1905. Application filed May 16, 1900. See also J. Erskine-Murray, A Handbook of Wireless Telegraphy, Chapt. 17, pp. 312-330, 1913 edition, published by Crosby Lockwood and Son, London, and Appleton & Company, New York.

This [Fig. 82] illustrates, on a larger scale, the earth. Here is my transmitter -- mine or anybody's transmitter -- because my system is the system of the day. The only difference is in the way I apply it. They, the radio engineers, want to apply my system one way; I want to apply it in another way.

This is the circuit energizing the antenna. As the vibratory energy flows, two things happen: There is electromagnetic energy radiated and a current passes into the earth. The first goes out in the form of rays, which have definite properties. These rays propagate with the velocity of light, 300,000 kilometers per second. This energy is exactly like a hot stove. If you will imagine that the cylinder antenna is hot -- and indeed it is heated by the current -- it would radiate out energy of exactly the same kind as it does now. If the system is applied in the sense I want to apply it, this energy is absolutely lost, in all cases most of it is lost. While this electromagnetic energy throbs, a current passes into the globe.

127

Now, there is a vast difference between these two, the electromagnetic and current energies. That energy which goes out in the form of rays, is, as I have indicated here [on the diagram of Fig. 82], unrecoverable, hopelessly lost. You can operate a little instrument by catching a billionth part of it but, except this, all goes out into space never to return. This other energy, however, of the current in the globe, is stored and completely recoverable. Theoretically, it does not take much effort to maintain the earth in electrical vibration. I have, in fact, worked out a plant of 10,000 horse-power which would operate with no bigger loss than 1 percent of the whole power applied; that is, with the exception of the frictional energy that is consumed in the rotation of the engines and the heating of the conductors, I would not lose more than 1 percent. In other words, if I have a 10,000 horsepower plant, it would take only 100 horsepower to keep the earth vibrating so long as there is no energy taken out at any other place.

There is another difference. The electromagnetic energy travels with the speed of light, but see how the current flows. At the first moment, this current propagates exactly like the shadow of the moon at the earth's surface. It starts with infinite velocity from that point, but its speed rapidly diminishes; it flows slower and slower until it reaches the equator, 6,000 miles from the transmitter. At that point, the current flows with the speed of light -- that is, 300,000 kilometers per second. But, if you consider the resultant current through the globe along the axis of symmetry of propagation, the resultant current flows continuously with the same velocity of light.

Whether this current passing through the center of the earth to the opposite side is real, or whether it is merely an effect of these surface currents, makes absolutely no difference. To understand the concept, one must imagine that the current from the transmitter flows straight to the opposite point of the globe.

There is where I answer the attacks which have been made on me. For instance, Dr. Pupin has ridiculed the Tesla system. He says,

"The energy goes only in all directions."

It does not. It goes only in one direction. He is deceived by the size and shape of the earth. Looking at the horizon, he imagines how the currents flow in all directions, but if he would only for a moment think that this earth is like a copper wire and the transmitter on the top of the same, he would immediately realize that the current only flows along the axis of the propagation.

The mode of propagation can be expressed by a very simple mathematical law, which is, the current at any point flows with a velocity proportionate to the cosecant of the angle which a radius from that point includes with the axis of symmetry of wave propagation. At the transmitter, the cosecant is infinite; therefore, the velocity is infinite. At a distance of 6,000 miles, the cosecant is unity; therefore, the velocity is equal to that of light. This law I have expressed in a patent by the statement that the projections of all zones on the axis of symmetry are of the same length, which means, in other words, as is known from rules of trigonometry, that the areas of all the zones must also be equal. It says that although the waves travel with different velocities from point to point, nevertheless each half wave always includes the same area. This is a simple law, not unlike the one which has been expressed by Kepler with reference to the areas swept over by the radii vectors.

I hope that I have been clear in this exposition — in bringing to your attention that what I show here is the system of the day, and is my system -- only the radio engineers use my apparatus to produce too much of this electromagnetic energy here, instead of concentrating all their attention on designing an apparatus which will impress a current upon the earth and not waste the power of the plant in an uneconomical process.

APPENDIX B
the Scalar wave

https://www.cia.gov/library/readingroom/docs/CIA-RDP96-00792R000500240001-6.pdf

SG1J   CONFIDENTIAL/NOFORN

From: DT-ACO ▮▮▮▮
To: DT (Dr. Vorona)

Subject: Scalar Waves

Ref: Verbal Request for Summary Statement on Scalar Waves

1. (C) Per reference, the writer will provide a summary below of his understanding of the nature of scalar waves. These are unconventional waves that are not necessarily a contradiction to Maxwell's equations (as some have suggested), but might represent an extension to Maxwell's understanding at the time. If realizable, the scalar wave could represent a new form of wave propagation that could penetrate sea water, resulting in a new method of submarine communications and possibly a new form of technology for ASW. Thus the potential applications are of high interest to the U.S. R&D Community and the Intelligence Community, particularly if some promise is shown to their realizability.

2. (C/NF) There is a community in the U.S. that believes that the scalar waves are realizable. In a recent conference sponsored by the IEEE these were openly discussed and a proceedings on the conference exists. The conference was dedicated to Nicola Tesla and his work, and the papers presented claimed some of Tesla's work used scalar wave concepts. Thus there is an implied "Tesla Connection" in all of this. ▮▮▮▮▮▮▮ SG1B

SG1B ▮▮▮▮▮▮▮▮▮▮▮▮▮▮▮▮▮▮▮▮▮▮▮▮▮▮▮▮▮

3. (U) The scalar wave, as the writer understands, is not an electromagnetic wave. An electromagnetic (EM) wave has both electric (E) fields and magnetic (B) fields and power flow in EM waves is by means of the Poynting vector, as follows:

$$\bar{S} = \bar{E} \times \bar{B} \quad \text{watts/m}^2$$

The energy per second crossing a unit area whose normal is oriented in the direction of $\bar{S}$ is the energy flow in the EM wave.

A scalar wave has no time varying B field. (In some cases it also has no E field.) Thus it has no energy propagated in the EM wave form. It must be recognized, however, that any vector could be added that could integrate to zero over a closed surface and the Poynting theorem still applies. Thus there is some ambiguity in even stating

$$\bar{S} = \bar{E} \times \bar{B}$$

is the total EM energy flow.

4. (U) The scalar wave could be accompanied by a vector potential $\bar{A}$ and $\bar{E}$ and yet $\bar{B}$ remain zero in the far field.

CONFIDENTIAL/NOFORN

From EM theory we can write as follows:

$$\bar{E} = -\nabla\phi - \frac{1}{c}\,\partial\bar{A}/\partial t$$
$$\bar{B} = \nabla \times \bar{A}$$

$\Big\}$ $Always \sim$.

In this case ϕ is the scalar (electric) potential and $\tilde{A}$ is the (magnetic) vector potential.

Maxwell's equations than predict

$$\nabla^2\phi - \frac{1}{c^2}\frac{\partial^2\phi}{\partial t^2} = 0 \qquad \text{(Scalar Potential Waves)}$$

$$\nabla^2\bar{A} - \frac{1}{c^2}\frac{\partial^2\bar{A}}{\partial t^2} = 0 \qquad \text{(Vector Potential Waves)}$$

A solution appears to exist for the special case of $\bar{E}=0$, $\bar{B}=0$, and $\nabla \times A=0$, for a new wave satisfying

$$\bar{A} = \nabla S$$
$$\phi = -\frac{1}{c}\frac{\partial S}{\partial t}$$

S then satisfies

$$\nabla^2 S - \frac{1}{c^2}\frac{\partial^2 S}{\partial t^2} = 0$$

Mathematically S is a "potential" with a wave equation, one that suggests propagation of this wave even through

$$\bar{E} = \bar{B} = 0$$

and the Poynting theorem indicates no EM power flow.

5. (U) From paragraph 4 above there is the suggestion of a solution to Maxwell's equations involving a scalar wave with potential S that can propagate without Poynting vector EM power flow. But the question arises as to where the energy is drawn from to sustain such a flow of energy. A vector that integrates to zero over a closed surface might be added in the theory, as suggested in para 3 above. Another is the possibility of drawing energy from the vacuum, assuming net energy could be drawn from "free space." Quantum mechanics allows random energy in free space but conventional EM theory has not allowed this to date. Random energy in free space that is built of force fields that sum to zero is a possible approach. If so, these might be a source of energy to drive the S waves drawn from "free space." A number of engineer/scientists in the community suggested in para 2 are now claiming this. A chief proponent of this is Lt Col Tom Bearden, who also lectured at the IEEE Tesla Symposium. He is known for his "Fer-de-Lance" briefing on "Soviet Scalar Weapons."

6. (U) In summary, scalar waves refer to non-EM waves with the potential for

CONFIDENTIAL/NOFORN

unconventional wave propagation. They appear to have some properties of soliton waves: they may not attenuate like EM waves do. Their existence is not proven, but if they exist their energy source is not clear. They have a quantum-mechanical flavor about them.

7. (U) If such scalar waves exist than they will be transformed via collective phenomena from microscopic waves to macroscopic waves, as in the case of Josephson junction theory (Cooper pair electron effects). They will also behave like longitudinal waves in a plasma and grow via propagation on a non-optical branch of the w-k space. They will result from collective phenomena, and (as in plasma waves) grow via energy supplied by the medium (free space, sea water, etc.). Should they exist, new vistas in wave propagation and long distance ranging/detection will result.

DT-ACO

12/22/87

SG1J

131

The author: **G. Martin Poole and his interest in Nikola Tesla's wireless power:**

Martin has been in the telecom business since 1965, and he has experienced many technical changes in his long career where he developed leading edge technology in telecom circuit test automation. The first automated DDS, analog data, and T1 carrier problem detection and analysis were developed under his leadership at Bell Atlantic (now Verizon), and at BELLCORE/Telcordia Technologies (now Ericson).

After retirement from telecom work, Martin started a software development business. He created TOT, a Task Oriented Testing guide for both experienced and inexperienced testers in the Telecom's Special Service Centers, the web-based guide was the first of its kind in 1998, and it is still used in 2018.

Martin's company, Wireless Power Technologies (**WPT**), USA, was formed in 2015. He and his associates began work at Microbotics Inc. in 2013. WPT has successfully re-developed Tesla's wireless power technology making use of modern components like Ferrite transformers, the C3 (*a three-plate capacitor used for reactive power communication*), and aerial receiver systems.

There are many firsts at WPT. The discovery of mechanical scalar waves in the electron medium, Longitudinal Electron Waves (LEWs), Electron Plasma (EP) fields, and the various versions of dynamic and reactive receivers to exploit these new discoveries.

Born out of this research was underwater light speed broadband with a range of hundreds of miles, coupled with underwater power delivery for drones and habitats to charge batteries or deliver direct power. The discovery that the longitudinal electron wave (LEW) is a true scalar wave with a mechanical compression and rarefaction characteristics.

There is nothing withheld from the reader about how this technology functions so that along with us, others can also move forward with Nikola Tesla's dream of a world power and communication system.

Professional work history:

- Verizon: Test system specialist and software engineering
- Bell Communication Research: Principal lead software development
- Telcordia Technologies and Ericson: Technical Director, Sr. Engineer, and Principal Director
- Spirent Communications: software development lead engineer
- Connect Data Wireless: CTO
- Wireless Power Technologies: CTO/Sr. engineer and scientist R&D for wireless power transport and broadband